basic

Luton

Luton

WITHDRAWN

Printed in the United Kingdom by MPG Books Ltd, Bodmin

Published by SMT, an imprint of Sanctuary Publishing Limited, Sanctuary House, 45-53 Sinclair Road, London W14 0NS, United Kingdom

www.sanctuarypublishing.com

Cover image courtesy of CORBIS

While the publishers have made every reasonable effort to trace the copyright owners for any or all of the photographs in this book, there may be some omissions of credits, for which we apologise.

ISBN: 1-86074-265-3

basic
Microphones

Paul White

smt

CONTENTS

INTRODUCTION

No matter how sophisticated computers or synthesisers become, the recording of a 'real' sound always starts with a microphone. The problem is that, unlike the human ear, there is no single microphone that is ideal for all jobs – microphones come in many types and sizes, and all are designed to handle a specific range of tasks. The problem is in deciding what microphone to choose for a particular application.

Having selected an appropriate microphone, there is still the question of how best to position it relative to the sound source in order to capture the desired sound. The aim of this book is to explain how the different types of microphones work, which types are best suited to which jobs, and how best to use them in a recording situation.

1 SOUND

Sound is created when a vibrating object (such as the body of an acoustic guitar or the head of a drum) causes the air about it to vibrate within the frequency range of human hearing. When this vibration reaches our ears it causes our eardrums to vibrate accordingly, and our brains perceive this as sound.

But even our ears have limitations, and the human hearing system can, at best, only detect air vibrations in the 20Hz–20kHz range (20 vibrations a second to 20,000), though there is variation from individual to individual. The upper limit of hearing deteriorates with age, and as a rule of thumb we lose around 1kHz for each decade of our age.

It is important to note that, although sound travels at around 1,100 feet per second, the air itself doesn't move. The way in which sound travels is often explained in textbooks by comparing it with the ripples formed in a pond after a stone has been thrown into the water. A cork placed on the water merely bobs up and down as the ripples pass by. It doesn't travel with them because the water itself isn't moving away from the point where the stone entered the water.

Sound behaves in much the same way, except that the ripples travel out from the source in all directions in a spherical manner, and as the sphere expands the sound energy gets weaker as the energy in the wavefront is spread over an increasingly large area. Even relatively loud sounds involve quite low amounts of acoustic energy unless you're very close to them, which all conspires to make the microphone's job surprisingly difficult. It has to convert this tiny amount of acoustic energy into a meaningful electrical signal that can later be amplified to a useful level.

Transducers

Any device designed to convert some form of physical energy to electrical energy, or vice versa, is known as a transducer. Examples include the pressure transducers used in electronic weighing machines; photo transducers, such as the photocells used in automatic cameras; the temperature transducers used in electronic thermometers and thermostats; and those transducers that convert motion into electricity.

Microphones fall into the latter category. However,

there is, as yet, no practical way of converting the vibration of air directly into electricity, so all commercially available microphones make use of some form of lightweight diaphragm. As the air vibrates the air pressure in the vicinity of the microphone diaphragm fluctuates, causing the diaphragm to move backwards and forwards over a small distance, closely following the vibrations of the original sound. To turn this tiny movement into an electrical signal there needs to be some system for measuring the movement of this diaphragm, and the exact nature of this system varies with the type of microphone. The overall result is an electrical signal that rises and falls in voltage to mirror the rise and fall in air pressure caused by the original sound.

It could be argued that, as we only need one set of ears to hear any type of sound at any level, it should be possible to make a single microphone that is suitable for all recording jobs. Some of the more expensive models come pretty close to this ideal, but the truth of the matter is that microphone technology isn't perfect, and in order to make a microphone perform particularly well in one area it's generally necessary to compromise its performance to some extent in another. For example, the frequency response of a microphone designed for very low background noise may not be as good as that of another, and it may not respond as accurately to off-axis sounds (those that don't occur directly in front of it). Also, remember that not everyone can afford the most exotic microphones, so it is important to be aware of the compromises that have been made in the name of economy and how they will affect the performance of the microphone in specific recording or sound-reinforcement situations.

Directionality

Not all microphones pick up sound in the same way, and the type you choose will depend on the task in hand. Some pick up sound efficiently regardless of the direction from which the sound is coming – in other words you don't have to point the microphone directly at the sound source because it can 'hear' equally well in all directions.

Some microphones may be designed to respond mainly to sounds approaching them from a single direction while others may pick up sounds from the both front and the rear but not from the sides. These basic directional characteristics are known as omnidirectional (all directions), cardioid (unidirectional – literally 'heart shaped') and figure-of-eight (mics which pick up from both front and rear but not from the sides).

It is possible to design a microphone incorporating two or more capsules, the outputs of which can be combined in a variety of different ways to give a selection of pickup patterns, and these will be discussed in more detail later in the book. First, though, it may be helpful to take a closer look at the individual pickup patterns.

Omnidirectional

If the diaphragm of a microphone is fixed across the end of a sealed, airtight cavity then the air pressure at the rear of the diaphragm will be essentially constant, while that on the side open to the air will vary depending on the sound reaching the microphone. Figure 1.1 shows how this works in practice. Because

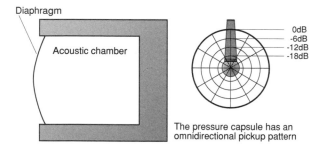

The pressure capsule has an omnidirectional pickup pattern

Figure 1.1: Basic pressure capsule

this type of microphone responds directly to changes in air pressure, it is known as a pressure microphone. Pressure changes will occur regardless of the direction from which the sound is coming, so the microphone will be omnidirectional.

In practice there must be a small hole to allow air to flow into the cavity in order to compensate for any changes in outside air pressure. Failure to provide this would mean that the microphone could also double as a barometer!

Because all microphones physically obstruct the soundfield they try to measure to some extent the omnidirectional response is often less than perfect. This usually results in the microphone being less sensitive to high frequencies at the sides and to the rear of the microphone than at the front. In theory, a perfect omnidirectional mic would be infinitely small so that it wouldn't interfere with the soundfield in any way, but as this simply isn't possible in practice then some compromises must unfortunately be accepted. However, these need not be serious.

Omnidirectional microphones do not exhibit the proximity effect inherent in cardioid and figure-of-eight designs. The proximity effect is explained in the section on cardioid microphones.

Figure-Of-Eight

The figure-of-eight pattern microphone uses a diaphragm that is open to the air on both sides. It is easy to see how this picks up sound from two directions, but why doesn't it pick up sound coming from the sides as well? The answer is quite simple when you stop and think about it. A sound arriving from the side will reach both sides of the diaphragm at the same time, and as a consequence the air pressure on each side of the diaphragm will always be at equal strength. If there is no difference in pressure then the diaphragm won't move and no

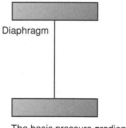

The basic pressure gradient capsule has a figure-of-eight response

Figure 1.2: Pressure gradient capsule

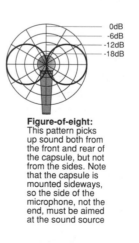

Figure-of-eight:
This pattern picks up sound both from the front and rear of the capsule, but not from the sides. Note that the capsule is mounted sideways, so the side of the microphone, not the end, must be aimed at the sound source

electrical output will be produced. Figure 1.2 shows the general principle involved with this. Because this type of microphone works on the difference in pressure between the front and the rear it is known as a pressure gradient microphone and consequently exhibits a proximity effect, which tends to result in an increase in the microphone's responsiveness to low-frequency sounds when the microphone is used very close to the source.

Cardioid

Another type of pressure gradient mic is the directional or cardioid mic, which is similar to the figure-of-eight mic except that a specially designed sound path is used to delay the sounds reaching the rear of the diaphragm. The design of this sound path is critical, and the top manufacturers keep their precise design details secret, but the outcome is that sounds arriving from the front of the microphone cause a pressure differential between the front and rear of the diaphragm, while sounds arriving from the rear and sides cause the pressure to be the same on both sides. The practical outcome is that the microphone is most sensitive to sounds arriving from directly in front and least sensitive to sounds arriving from the rear. In reality, sounds arriving from the sides are still picked up to some extent, although less efficiently

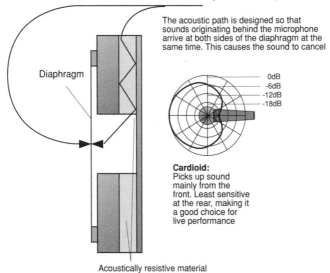

Off-axis sound from directly behind the capsule

The acoustic path is designed so that sounds originating behind the microphone arrive at both sides of the diaphragm at the same time. This causes the sound to cancel

Diaphragm

0dB
-6dB
-12dB
-18dB

Cardioid:
Picks up sound mainly from the front. Least sensitive at the rear, making it a good choice for live performance

Acoustically resistive material

The pressure gradient capsule may be modified to produce a cardioid (or unidirectional) response

Figure 1.3: Directional microphone capsule

than those from the front, giving rise to the characteristic heart-shaped or cardioid pickup pattern from which this type of microphone derives its name. A simplified diagram of a unidirectional microphone is shown in Figure 1.3.

A microphone that is even more strongly directional is called a supercardioid or hypercardioid. An extreme hypercardioid with a very narrow pickup pattern can be produced by fixing the microphone capsule to the rear of a complex interference tube structure, and these devices are known as shotgun mics because of their appearance. They are often used for spot-miking during stage shows, where it isn't possible to get the microphone close to the performer without spoiling the visuals, and they are also useful for recording wildlife and some forms of surveillance work. Their design makes them inefficient at low frequencies, so they are seldom used in music studios. All cardioid microphones exhibit the proximity effect, as explained below.

The Proximity Effect

Pressure gradient microphones (cardioids and figure-of-eights) both exhibit the proximity effect, which causes a boosting of low frequencies when the sound source is very close to the microphone. This problem arises because the path lengths for sounds arriving at the front and rear of the diaphragm are not quite the same. If the distance between the mic and source is very small then the difference between these two acoustic paths is proportionally more significant, and the phase differences in the signals give rise to the characteristic low-frequency boost. Typically, this happens at mic/source distances of a couple of inches or less, and is used to advantage by some live vocalists to modify their performances.

Attributes

These different directional patterns allow the microphones to be used in a variety of different situations. The unidirectional (cardioid), for example, is used where unwanted off-axis sound, such as room reverberation or spill from other instruments, needs to be kept at a minimum. This is often the case in recording studios, where several musicians may be playing close together, and also when miking a drum kit, where the close proximity of the individual drums is an even greater problem. But it should be borne in mind that any spill that does exist will be coloured by the off-axis frequency response of such microphones. In other words, if the microphone loses some top end when picking up sounds from the side or from the rear then any spill from off-axis sources will tend to sound less bright than it really is.

Separation is particularly important in live situations, and so cardioid mics are extensively used with touring PA systems, as well as for recording drum kits or other instruments that are physically close to each other.

The figure-of-eight-pattern microphone is usually used only for specialist applications such as stereophonic miking, which will be discussed in detail later in this book. They were also popular at one time for using with live backing vocals, as two singers could share one mic – one at the front and the other at the rear.

Omnidirectional microphones have an inherently more natural sound than cardioids because there is no need for any mechanical porting to modify their directional characteristics. This means that even off-axis sounds are reproduced reasonably faithfully. Other advantages include a higher resistance to handling noise than cardioid mics, immunity from the proximity effect and a greater capacity for handling high levels of sound pressure.

They are often used at conferences or in the production of radio programmes, to pick up a group of speakers seated around a table, but are also widely used in serious music recording (both for solo instrument and vocal miking) and for specialist stereo

basic Microphones

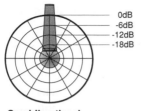

0dB
-6dB
-12dB
-18dB

Omnidirectional:
Picks up sound equally from all directions. Used mainly for recording or for picking up multiple sound sources at the same time

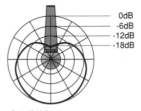

0dB
-6dB
-12dB
-18dB

Cardioid:
Picks up sound mainly from the front. Least sensitive at the rear, making it a good choice for live performance

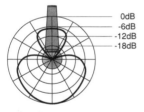

0dB
-6dB
-12dB
-18dB

Hypercardioid:
Sometimes called a Supercardioid, this is a narrower pattern than the cardioid, but is more sensitive to sounds coming directly from behind. Care should be taken to place monitors in the mic's dead zone

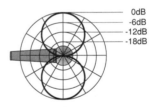

0dB
-6dB
-12dB
-18dB

Figure-of-eight:
This pattern picks up sound both from the front and rear of the capsule, but not from the sides. Note that the capsule is mounted sideways so that the side of the microphone, not the end, must be aimed at the sound source

Figure 1.4: Common pickup patterns

work. Figure 1.4 shows response plots for each of the common pickup patterns produced by studio microphones.

Even in situations in which separation is a prime requirement it can be argued that omnis give a more natural result, as any off-axis spill will be recorded faithfully rather than suffer distortion from an imperfect off-axis response, which would be the problem with most cardioid mics. Although the hypercardioid mic is designed to exclude as much off-axis sound as possible it is interesting to note that, if you position an omni mic at between half and two thirds the distance of a cardioid mic, the amount of spill will be similar.

Construction

The transducer that converts sound into electricity is built into the part of the microphone known as the capsule. This is usually mounted behind a protective grille, and it is commonplace for microphones also to have some form of handle or main body in which to house any electronics, transformers or connectors that might be needed.

Whatever the microphone's working principle, the actual size of the diaphragm is vital to the way in which it picks up sounds. This will be explained late in this book in the section covering the dimensions of capsules.

Studio and live sound microphones tend to use a flexible shock mounting in order to minimise the level of handling noise which reaches the capsule, and the protective grille very often incorporates a fine-mesh wind shield which helps to reduce the level of popping on vocals. As we shall see later, these shields are of very limited use, and so some form of external pop shield is usually necessary for discerning work. Fortunately, these can be improvised for negligible cost.

2 DYNAMIC MICROPHONES

The dynamic (or moving-coil) microphone works on much the same principle as the generators that provide our mains electricity, albeit on a much smaller scale. A light circular diaphragm, usually made from a thin plastic film, is attached to a very fine coil of wire, which in turn fits into a gap in a permanent magnet in such a way that it can move freely between the north and south poles. In this respect the dynamic microphone assembly looks very much like that of a moving-coil loudspeaker, except that the speaker works the other way round: it turns electrical signals back into sound.

When the diaphragm moves back and forth in response to a sound, the attached coil moves within the magnetic field, which generates an electric current in the wire of the coil. This current is very small but can easily be amplified to a useful level by a pre-amplifier, such as the input circuitry of a mixing console. Figure 2.1 over the page shows the construction of a typical moving-coil microphone.

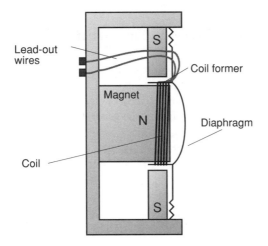

Figure 2.1: Moving-coil microphone

Pros And Cons

Dynamic microphones have several advantages over other types of microphone. They are relatively inexpensive to manufacture and are very rugged, which means that they can be used live as well as in the studio. They can also tolerate extremely high levels of sound pressure and require no power supply, as there is no electronic circuitry in the microphone itself.

However, dynamic mics also have their disadvantages. Firstly, the movement of the diaphragm is restricted to some extent by the mass of the coil attached to it. The faster the diaphragm tries to move the more the inertia of the coil impedes it, and high-frequency efficiency consequently suffers. In practice a conventional dynamic microphone will work effectively up to around 16kHz, but above that the efficiency tends to fall significantly.

Some recent improvements have been achieved by using a new material for the magnetic structure: neodymium. This produces a more intense magnetic field, allowing the use of a shorter coil, and the smaller mass of the coil allows the diaphragm to move more freely at higher frequencies. There is some disagreement between manufacturers as to how great the advantages of neodymium mics really are, but there are several models on the market which boast a frequency response exceeding 20kHz.

Dynamic microphones are further disadvantaged in that they produce a relatively small output signal, which needs a lot of amplification to make it usable. This is no problem if the sounds being picked up are moderately loud and close to the microphone, but soft or distant sounds often require so much amplification that the results are unacceptably noisy. For this reason, the dynamic mic is rarely used to record instruments like the acoustic guitar because it's difficult to get the mic close enough to pick up a reasonable amount of sound without compromising the tone of the instrument.

Capsule Dimensions

There is a common misconception that dynamic microphones themselves can't actually generate noise because they have no active circuitry. The truth is that any circuit carrying electrical resistance produces electrical noise, and the higher the resistance the higher the noise. Furthermore, the impact of individual air molecules on the diaphragm of any type of microphone generates noise, in much the same way that individual oxide particles on recording tape contribute to tape hiss, though it is true that this is usually only significant with very small capsules.

A larger diaphragm is more efficient at eliminating noise as the sound picture is built up from the impact of more air molecules. This gives a statistically better result, in much the same way as a wider format of recording tape or a faster speed of tape gives lower tape hiss on a tape machine, although large diaphragms cause compromises in other areas.

Sounds approaching the microphone head-on will reach all parts of the diaphragm at more or less the same time, but sounds coming in at an angle will reach one side of the diaphragm before the other, so some of the higher frequencies will combine out of phase, causing a deterioration in high-frequency response for sounds arriving off axis. Furthermore, the larger the diaphragm the greater the handling noise of the microphone. If the microphone is knocked or moved suddenly, the higher inertia of the diaphragm means that it tries to stay where it is, so it is in effect in motion relative to the rest of the capsule. Movement between the diaphragm and the magnetic assembly is the mechanism which creates an output signal, which is why the sudden movement of a microphone can elicit a thump!

Another drawback of a large diameter of diaphragm is the microphone's physical size, which may interfere with the very soundfield that it is trying to capture, particularly at high frequencies where the wavelength of sound is shortest. As in most areas of design, the diameter of the diaphragm is a calculated compromise and differs from one model to the next, depending on its application. On the positive side, the lower resonant frequency of a large-diameter capsule makes it more suitable for use with bass instruments.

basic Microphones

Dynamic mics are generally available in omnidirectional and cardioid versions, and cardioid mics break down further into wide-pattern cardioids, normal cardioids and hypercardioids. Of the three, the hypercardioid is the most directional.

Because of the physical construction of dynamic microphone capsules, it is generally considered impractical to use more than one diaphragm, which means that each model is usually designed to offer only one fixed pickup pattern.

3 RIBBON MICROPHONES

The ribbon microphone works on the same electrical principle as the dynamic microphone, except that a thin conductive ribbon, usually made from aluminium, plays the role of both the diaphragm and the coil. This arrangement is the electrical equivalent of a dynamic microphone, with just a single turn coil so that only a tiny current flows through the ribbon in response to a sound. Because the electrical output from the ribbon is so low, a transformer is incorporated into the design to bring the voltage up to a usable level. As with any microphone, it is also important that the ribbon assembly is well screened to prevent outside interference from polluting the signal. Figure 3.1 shows how such a microphone is constructed.

A well-designed ribbon microphone can have a frequency response rising to 20kHz, and this response can be made very flat, though many roll off below 20kHz. The double-sided nature of the ribbon construction produces a pressure-gradient microphone with a figure-of-eight pickup pattern, though some

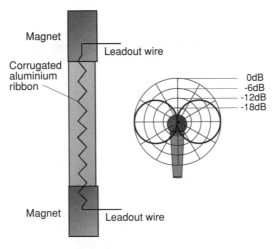

The ribbon mic has a figure-of-eight response

Figure 3.1: Ribbon microphone

modified designs are available with more directional characteristics.

Some people argue that, because ribbon microphones are no more sensitive than moving-coil microphones, and because the frequency response of many commercial models is no better than that of a good moving-coil model, they offer no real advantage over more conventional dynamic designs that are invariably cheaper. Additionally, older ribbon microphones had the severe disadvantage of being physically very fragile. Not only could rough handling damage the ribbon but also, over a long period, condensation from a singer's breath could corrode the thin aluminium ribbon, eventually causing it to fail. But, as in all things, technology has since progressed, and modern ribbon mics are relatively tough. There have also been new design innovations, such as the printed ribbon, whereby a metallic film is deposited on a light plastic membrane, combining some of the advantages of a traditional ribbon microphone with those of a capacitor model. This particular design approach is said to make the ribbon more durable as well as extended its frequency response.

Despite the fact that ribbon mics may seem like old technology, many professionals find that they produce the best subjective sound in particular applications, especially when recording classical string sections. For this reason, certain vintage models are highly valued. By some quirk of physics, they combine the detail and subtlety of a capacitor microphone with the more rounded, smoother sound of a traditional dynamic mic.

4 CAPACITOR MICROPHONES

A capacitor – or a condenser, as it is sometimes called – comprises a pair of parallel metal plates separated by an insulator. Its claim to fame lies in the fact that it can store an electrical charge, and for those interested in technicalities the relevant formula is $Q = CV$, where Q is the electrical charge in coulombs, C is the capacity in farads and V is the voltage across the two plates. Capacitors are used in all kinds of electronic circuitry, but the principles of capacitance also make it possible to build extremely good microphones.

If the capacitance is varied by altering the distance between the two plates of a charged capacitor then the voltage across the plates will also change, as a result of which a current will flow into or out of the capacitor through the resistor connecting it to the power supply. A signal can be obtained by monitoring the voltage across this resistor by means of a high-impedance pre-amp.

A capacitor microphone comprises two such plates: one solid, fixed metal plate and one very thin, flexible plastic diaphragm onto which has been deposited an extremely thin metal coating to make it electrically conductive. If the plates are then electrically charged, any movement of the diaphragm caused by vibrations in the air will cause the capacitance to change accordingly, and this change is then translated into a voltage and amplified to produce an audio output.

Charge

In addition to the capacitor capsule, some means of providing an electrical charge to the diaphragm and back-plate is needed. The charge is usually provided either via the 48v phantom power source in the mixing console or mic pre-amp or from a separate phantom power supply. Phantom powering will be discussed more fully later in this book.

In order to amplify the voltage changes on the capsule without allowing the electrical charge to leak away, a pre-amp with a very high input impedance is used. In early designs these employed valves, but now FETs (Field Effect Transistors) are more common. Valve designs are still popular, however, and well-preserved older models change hands for considerable sums of money. There are also numerous new and reissued valve designs, and it's thought that the reason they sound so distinctive is that valve circuitry tends to subtly colour the sound in a way that many engineers and producers find appealing.

Not only are they cheaper but FETs also don't require the heavy heater currents required by valves, and so the pre-amp can also be run from the phantom powering system – the microphone is in effect completely self-contained. On the other hand, valve microphones

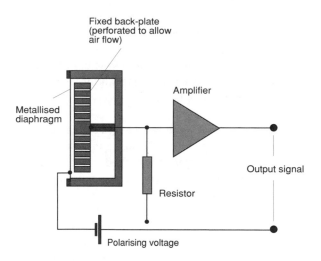

Figure 4.1: Internal construction of a typical capacitor microphone

require a complex and often costly power supply which has to be capable of supplying not only the capsule polarising voltage but also the high tension and heater supplies required by the valve itself. Figure 4.1 shows the internal construction of a typical solid-state capacitor microphone. Note that the fixed back-plate is perforated so that air can pass freely through it.

The capacitor microphone seems rather complicated when compared with the moving coil or dynamic microphone, and because many of the manufacturing stages still need to be carried out by hand this is often reflected in the cost. Even so, the system offers several real advantages over other types of microphone, and the majority of serious recording work is carried out with capacitor microphones of one type or another. Fortunately, as private recording studios have become more popular and the cost of using them has dropped, cost-effective capacitor microphones have now become available, many of which perform exceptionally well.

Advantages

Probably the most important advantage of the design of the capacitor microphone capsule is that the metal coating on the plastic diaphragm can be made just a few microns thick, which means that it is very light. Less weight consequently means less inertia, which in turn means that the diaphragm can respond more effectively to higher frequencies than the dynamic microphone. In conjunction with a high-quality on-board pre-amp, capacitor mics offer the best noise performance and the highest sensitivity of any studio microphone. Furthermore, their frequency response can easily exceed the range of human hearing at both the high and low ends of the audio spectrum.

Capacitor microphones can be made with virtually any response pattern, and systems using two or more diaphragms are often found in studios because they enable the microphone to be switched over a range of different response characteristics.

Refinements
Current capacitor microphone designs are rugged enough to be used in live situations as well as in the studio, but many suffer from loss in sensitivity if used in a very humid environment. Even in an air-conditioned studio, the moisture from a singer's breath can cause problems. The reason for this is fairly simple: the moisture forms a conductive path that allows the electrical charge on the capsule to leak away.

RF Capacitors
European microphone manufacturer Sennheiser have

resurrected a system first used in the very early days of capacitor mics and refined it to a point at which its advantages are clearly significant. Their method is to bias the capsule with a rapidly alternating voltage rather than with a fixed electrical charge. This bias is provided by an oscillator running at around 8mHz and gives the technique its name: RF (Radio Frequency). In effect, the audio signal is superimposed on this high-frequency oscillation, just like the music on an FM radio signal, and circuitry within the body of the microphone extracts the audio signal before feeding it out as normal.

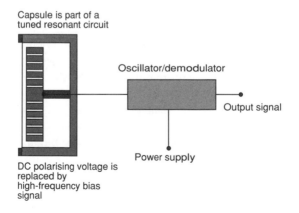

Capsule is part of a
tuned resonant circuit

Oscillator/demodulator

Output signal

DC polarising voltage is
replaced by
high-frequency bias
signal

Power supply

Figure 4.2: RF microphone principle

A high-frequency capsule working on this principle has an inherently high impedance, so normal levels of moisture in the air have no significant effect on its performance. Early RF designs were unstable, noisy and susceptible to some forms of RF interference, but technology has moved a long way since then and the modern version performs very reliably. Figure 4.2 shows a typical RF microphone system.

Symmetry

One limitation of the basic capacitor microphone is that the capsule is not symmetrical, either electrically or acoustically. When the diaphragm moves towards the back-plate, the acoustic impedance is different to that when it is moving away because the spacing is slightly different depending on whether the diaphragm is bending inwards or outwards. The electrostatic attraction exerted on the diaphragm by virtue of the electrical field between it and the back-plate is asymmetrical for the same reason.

One solution to this problem involves building a microphone with two perforated plates, one in front of the diaphragm and one behind it. These are both part of the electrical circuit, and so the capsule is both electrically and acoustically symmetrical within the limits imposed by manufacturing accuracy. The symmetrical construction also reduces intermodulation distortion by a significant degree, though different manufacturers have different methods of optimising the performance of their capsules.

Intermodulation Distortion

Briefly, intermodulation distortion manifests itself in a system that is in some way non-linear – an asymmetrical capsule design, for example. The outcome is that, when two different frequencies are fed into the mic at the same time, the output contains small levels of the sum and differences of these two frequencies. For example, a tone of 2kHz and a tone of 3kHz would have intermodulation products at 1kHz and 5kHz. Even though the amount may be very small, it still affects the overall sound quality and can make the output from an otherwise good microphone sound instead harsh and confused. Improved capsule linearity reduces this effect.

Electrets

Electret mics work on a very similar principle to the capacitor and have been around in one form or another for several years. The main difference is that, with electret mics, the electrical charge on the diaphragm is not provided by a power supply but is built in at manufacture by a process involving heat and strong magnetic fields. Exactly how this is achieved is an industrial secret, but early models had the charge-carrying elements built into the insulating material that formed the diaphragm. This was made of a highly insulating plastic, to ensure that the charge remained intact for many years. An FET pre-amp was still needed to process the signal from the capsule, but this could be run from batteries contained in the mic's handle.

However, the real problem was fact that the very process of making a charged diaphragm meant that the diaphragm had to be thicker, and consequently heavier, than that of a conventional capacitor mic, and, as discussed in the section on dynamic mics, this resulted in a loss of sensitivity and high-end frequency response. Even so, they could be built very cheaply and needed no external power supply or phantom powering (most ran from a 1.5v battery), and were used extensively in domestic audio equipment such as cassette recorders.

Back-Electrets

Improvements in diaphragm technology enabled better mics to be built using the electret principle, but the most significant improvement turned out to be the invention of the back-electret capsule. This works on much the same principle, but with one very important difference: the material carrying the permanent electrical charge is fixed to the stationary back-plate, allowing the moving diaphragm to be made of exactly the same material as that used in a true capacitor microphone.

This back-electret technology has enabled a new range of cost-effective mics to be produced which offer virtually all of the benefits of true capacitor mics at around the cost of an average dynamic mic. The idea may seem obvious, but the manufacturing process took a lot of development as it proved to be very difficult to find a reliable way to bond the charged material to the back-plate. Figure 4.3 shows the

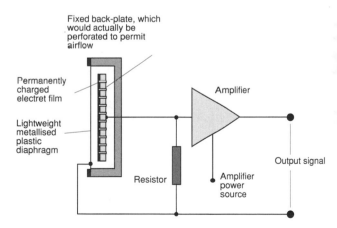

Figure 4.3: Construction of a back-electret microphone

construction of a back-electret microphone. There are back-electret microphones available which can run on batteries as well as on phantom power, but as a rule the higher phantom-power voltage results in more headroom and a better dynamic range than can be achieved from batteries.

Multipattern Microphones

Though it is feasible (and indeed has been accomplished) to build a variable pattern microphone capsule using dynamic or ribbon transducers, the capacitor system lends itself far more readily to this application. This is partly because the capacitor capsule is conceptually both compact and mechanically simple and also, importantly, because the output of a capacitor capsule can be governed simply by varying the polarising voltage between the back-plate and the diaphragm. Unfortunately, the cheaper electret principle doesn't rely on an external polarising voltage and so can't be controlled in the same way.

Many popular studio capacitor mics offer switchable response patterns, the Neumann U87 and AKG 414 being popular examples. The logic behind the variable pattern design is obvious: for the price of one microphone you get a model that can be used in a wide variety of recording situations.

Dual Cardioids

By combining two cardioid elements in a back-to-back fashion, all of the common response patterns can be produced by mixing the outputs from the two capsules in different amounts and phases. In physical terms this might require a dual diaphragm capsule in which the polarising voltage on one of the elements can be switched to vary the gain and phase of its output. Once the outputs from the two parts of the capsule are combined you should find that the desired pattern is produced.

If two cardioid patterns are combined equally and in phase then an omnidirectional response is produced, while switching off one of the cardioids will obviously result in a straightforward cardioid response from the other. On the other hand, if the two cardioids are combined equally and out of phase then this will result in the classical figure-of-eight pattern being produced. In addition to these three important basic patterns, varying the output level from one of the capsules can create all of the patterns in between, such as the wide cardioid and hypercardioid responses. Figure 4.4 shows the five most useful patterns which can be created in this way.

Another approach is to combine the outputs from a figure-of-eight mic and an omni to produce a cardioid response. The advantage of this arrangement is that no rear porting

basic Microphones

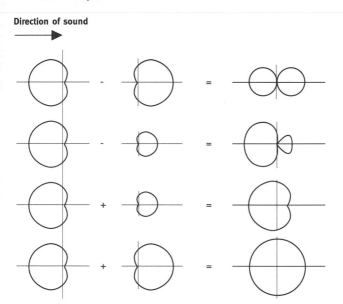

Figure 4.4: Patterns generated by two back-to-back cardioid capsules

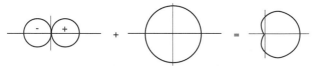

Figure 4.5: Production of a cardioid response from the combination of an omni and a figure-of-eight element

is required. In theory this means that a better-sounding cardioid can be produced, and by simply switching off one element or the other, both omnidirectional and figure-of-eight responses can also be produced. Figure 4.5 illustrates how this can be achieved.

All of the gain and phase control needed to produce these pattern changes is obtained by simply varying the polarising voltage to one side of the capsule. For example, if a phase reversal is necessary then the polarising voltage is simply reversed. Because the voltage which is being switched is DC at a relatively high level (60v or so) then the switching can be carried out remotely, if need be, with no detrimental effect to the resulting output signal of the microphone. Figure 4.6 shows the switching arrangement for a multipattern mic.

Sensitivity

The sensitivity of a microphone is calculated by measuring the electrical output for a given level of sound pressure. A high sensitivity is an obvious advantage in this respect as it helps to provide a good signal-to-noise performance – if the microphone produces a large signal then less amplification is necessary at the mixer end of the system, and it's well known that the noise produced by an amplifier increases as its gain increases.

basic Microphones

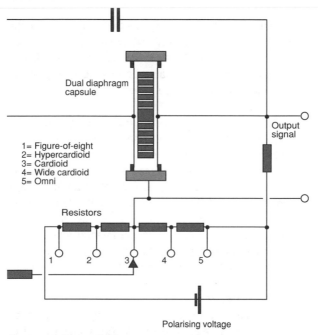

Dual diaphragm
capsule

Output
signal

1= Figure-of-eight
2= Hypercardioid
3= Cardioid
4= Wide cardioid
5= Omni

Resistors

1   2   3   4   5

Polarising voltage

Figure 4.6: The switching arrangement for a multipattern microphone

The conventional method of measuring is to present the microphone with a constant SPL (Sound Pressure Level) of 1 pascal or 10 microbars and then to measure the output voltage of the microphone with a high-impedance measuring system to avoid loading the microphone in any way. This is known as open-circuit measurement, and a typical studio microphone might have an output of 8–10mv/pa.

Research has indicated that the threshold of hearing (for a 1kHz tone) is around 20×10^4 pascals, so if this threshold is used as a 0dB reference we can express the SPL on a scale of decibels for convenience. On this scale, 1 pascal is equivalent to an SPL of 94dB.

Manufacturers often state the electrical noise generated by a particular microphone as an equivalent SPL – in other words, the level of an external sound that would produce the same signal level at the output of a perfectly noise-free microphone. This may be measured from 20Hz–20kHz using a flat (unweighted) response meter, or it may be measured using the so-called 'A-weighting' system, which has a frequency characteristic designed to compensate for the sensitivity of the human ear to different frequencies. A-weighted figures generally look a decibel or two more favourable than unweighted figures.

If we then subtract the equivalent noise SPL from the maximum SPL that the microphone can tolerate before distortion becomes unacceptably high then we are left with the useful dynamic range. An example might be a microphone with a noise SPL equivalent of 15dB and an upper overload threshold of 135dB. Subtracting these gives a dynamic range of 120dB. When you consider that the dynamic range of 16-bit digital recording is only 96dB or so, this figure looks very favourable. Of course, for quieter sounds that don't approach the mic's upper threshold, the noise performance will be correspondingly worse. This is an almost exact analogy of what happens with analogue recording tape, whereby the quieter the input signal the louder the comparative background hiss. For this reason quiet sound sources should always be recorded using the most sensitive microphones available if the best signal-to-noise performance is to be preserved.

When the sensitivity of a microphone is measured, it's generally done so over a range of frequencies and at a range of angles around the microphone so that the way in which it behaves towards off-axis sounds can be established. This may be achieved by mounting the microphone being tested on a turntable in closely controlled acoustic conditions such as an anechoic chamber. A calibrated level of tone is provided by a loudspeaker system at a fixed distance from the turntable and the output of the microphone connected to a plotting device incorporating a microphone pre-

Capacitor Microphones

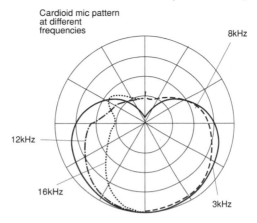

Cardioid mic pattern at different frequencies

8kHz

12kHz

16kHz

3kHz

Note how the polar pattern narrows at higher frequencies

Figure 4.7: Polar plot and graph showing a typical studio microphone's frequency response

At first glance, a dynamic vocal mic's frequency response may appear to be anything but flat, but these characteristics are built in for good reasons

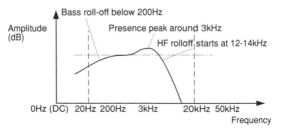

Bass roll-off below 200Hz

Presence peak around 3kHz

HF rolloff starts at 12-14kHz

Amplitude (dB)

0Hz (DC) 20Hz 200Hz 3kHz 20kHz 50kHz

Frequency

amplifier and level-measuring circuit. The recording paper is then rotated to follow the microphone's position while the position of a moving pen follows the output level from the microphone. In this way the sensitivity may be plotted over a full 360º. This test is then repeated using various frequencies throughout the audio range, and the results are plotted out on a single circular diagram to form the familiar polar diagram response found in microphone documentation. Figure 4.7 shows a typical polar plot and graph showing a typical studio mic's frequency response over a range of frequencies.

5 ELECTRICAL CONSIDERATIONS

Balancing

Microphones used in domestic equipment or for low-budget musical applications are connected to the amplification system by a standard screened cable. This cable comprises a single insulated central core surrounded by a screen woven from fine wire. Outside this screen is a rubber or synthetic protective coating. The reason that this type of cable is used rather than, for example, ordinary twin flex is that the output signal from a microphone is at a very low voltage and so is liable to be corrupted by electrical interference. Such interference is generated by equipment incorporating unshielded mains transformers and mains power wiring, and by any electrical circuit that switches or interrupts the current, such as thermostats, light dimmers and computer monitors. Radio signals can also cause problems, often producing what is known as RF (Radio Frequency) interference.

The screened cable offers some protection against interference but is by no means totally effective. The idea is that any radiated interference is intercepted by the outer screen, which is connected to electrical earth so that all of the interference drains away to earth. In practice, however, a significant proportion of the interfering signal still ends up superimposed on the wanted signal from the microphone. Part of the problem is that the screen forms the return path for the electrical circuit as well as providing the screening.

A more satisfactory approach is to use a balanced system. Figure 5.1 shows that this system uses two central cores in the cable, and these are again surrounded by a protective screen, though the screen does not form part of the signal path. Both the microphone and the pre-amplifier to which it is connected must be designed for balanced operation in order for this system to work.

Split Phase

Inside the microphone the output signal is split into two opposite phases, usually by means of a transformer or an electronic output balancing stage. These two phases are known as positive and negative, or hot and cold, and are fed through the two central cores of the microphone cable. The screen is connected to ground at the mixer end and to the body of the microphone at the mic end, but it doesn't carry any of the mic output signal.

Unbalanced

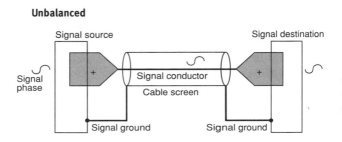

Balanced

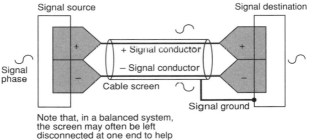

Note that, in a balanced system, the screen may often be left disconnected at one end to help prevent ground loops

Figure 5.1: Balanced wiring system

At the mixer or pre-amp end of the system, another transformer or electronically balanced input stage inverts one of the phases before adding its voltage to the other. Because the signals are once again in phase at this point they combine rather than subtract, leaving us with a usable signal.

Interference is treated rather differently because it affects both inner cores of the cable in the same way: an interference signal on the hot (positive) lead will also be present in the cold (negative) lead, as both are physically very close. When this noise signal encounters the balanced amplifier input one of the phases is inverted so that, when the signals are combined, the two equal noise signals are out of phase with each other and cancel out. The result is that the wanted signal is preserved but virtually of all the interference disappears. I say virtually here because, for perfect operation, the system relies on the balanced amplifier or input transformer to be perfectly symmetrical and effective at all frequencies. In reality, component tolerances mean that the circuit doesn't behave symmetrically, and most circuits are also less effective in cancelling interference at higher frequencies. In spec sheets, this capacity for cancelling noise is described as 'common mode rejection', and is expressed in decibels, usually specified at a single frequency, such as 1kHz.

Because balancing isn't 100% effective in rejecting unwanted interference, it's still wise to take precautions against interference by keeping cable runs as short as is practical and not routing cables close to mains cables or equipment containing transformers.

Impedance

Microphones come in two impedance types: high and low. High-impedance mics, typically 5–10kohms, are often used in domestic equipment and budget musical equipment because they have a relatively high output voltage compared to low-impedance types. This means that the pre-amps needn't be so electrically complex, as they don't need to provide as much gain.

This sounds to be an ideal system, but there are two major limitations involved: firstly, the higher impedance means that the signal is more susceptible to electromagnetic interference; and secondly, the higher the impedance the more the signal is then affected by cable capacitance.

Because the central core of a screened cable is in close proximity to the outer screen, a length of cable acts as a capacitor, and therefore the longer the cable the higher the capacitance. Different cables have different capacitance values, and these are usually specified in picofarads per metre. This capacitance attenuates the higher frequencies (the impedance of a capacitor falls as the frequency increases), and, because a typical microphone cable may have a capacitance in the order of 100pf per metre, the effect is significant. For this reason, high-impedance microphones can be used only with relatively short cable runs (ten or 20 feet is fine), but bear in mind that the longer the length the greater the treble loss and the higher the susceptibility to interference. Cable capacitance can also result in handling noise – any distortion of the cable that results in a variation in the distance between the screen and cable core can cause the cable to act like a capacitor microphone, introducing noise on the wanted signal.

Low-impedance microphones are always used for professional applications, and these have impedances of 250ohms or less. The microphone pre-amp will usually have an impedance of five to ten times that amount because we are concerned with transferring the maximum signal voltage to the pre-amp, not the maximum power. Also, by making the input impedance higher than that of the microphone, the effect of cable impedance is less significant than it would be if the input and output impedances were identical. With this system it's possible to use cable lengths of around 50m, which will incur only a decibel or so of high-frequency loss at the 2okHz mark. With both low- and high-impedance mics, the effect of cable resistance is negligible for any practical length of cable.

Cable Types

It's important to select a microphone cable with a low capacitance, especially if long runs are to be used. A woven, braided screen generally offers better protection against interference than the type in which the screen conductor is merely wrapped around the central cable, and indeed this latter type should be avoided in any instance, except possibly when using short runs of line-level signal. The most important reason for this is that the helically wound screen of a wrapped cable can open up at the point at which the cable is bent, thus compromising its effectiveness against interference.

There's another type of cable which uses a conductive plastic screen that may be used to make up microphone leads, and although it doesn't offer quite the same level of screening as the woven screen type it is still good enough for short- to medium-length cables, and has the distinct advantages of being resistant to kinking and having a low handling noise. It is also easy to use, as an uninsulated wire runs alongside the conductive plastic screen to form the screen connection. This saves time in stripping back and preparing the screen, and the only special precaution you need to make is to ensure that the conductive plastic screen is stripped back far enough to prevent it from shorting out on any part of the connectors to which it is soldered.

Connectors

Unbalanced microphones are often supplied with a fixed lead or a lead that is attached to the microphone using a non-standard, proprietary connector. The other end may be soldered to a standard quarter-inch jack or an XLR connector wired for unbalanced use by linking the earth and cold pins.

Balanced mics tend to rely on a separate XLR cable which plugs into the microphone handle, though some budget systems use stereo quarter-inch jacks wired tip hot, ring cold. A standard mic cable uses a female XLR three-pin connector at the mic end and a male three-pin XLR at the other. The three pins of the XLR plugs and sockets are numbered, and the easy way to remember the connections is to think: one, two, three = XLR, where X is earth, L is live (or hot) and R is return (or cold). In other words, the screen connects to pin one, the hot to pin two and the cold to pin three. This applies to all balanced systems, though some American companies still build equipment on which pin three is hot and pin two is cold. Figure 5.2

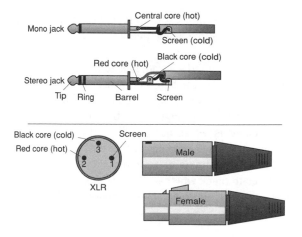

Figure 5.2: Wiring details for jacks and XLRs

shows the wiring for both balanced and unbalanced jacks and XLRs.

Phase Problems

If a microphone cable is wired with pins two and three swapped over it will still work as normal, until it is used in a multimic setup. A properly wired mic translates a positive increase in air pressure at the diaphragm into a positive voltage at the hot terminal, but crossed wiring anywhere along the chain will cause a positive pressure to produce a negative voltage.

If two correctly wired mics are placed in close proximity and their outputs combined in a mixing console then the resulting signal level will be around 6dB higher than the output from a single mic, but if one of these microphones is wired out of phase then the output signals will largely cancel each other out. This effect can be used as the basis of a simple test which can be employed to check whether all of your mics are wired in phase – even if all your cables are wired correctly, you could still have a mic which is wrongly wired internally.

To run the test, set up two mics side by side, one of which you know is correct, and place a loudspeaker a little way in front of them. Feed the speaker from a test oscillator (around 200Hz) or from a synth pitched at around one octave below middle A and set up the gain controls on the mixer (one mic at a time) so that both mics give the same VU meter reading. When both faders for both microphones are up the combined signal should give an increased meter reading. If the reading drops then the chances are there's a phasing problem.

The actual test frequency isn't actually critical, but it should be kept at a fairly low level to ensure that the

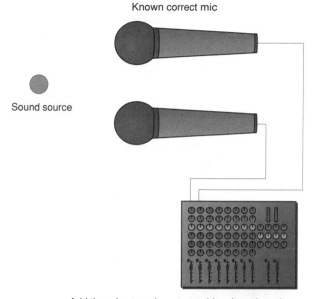

Known correct mic

Sound source

Add the mics together at equal levels and monitor
the level on the meters. If switching off one mic
causes an increased reading, the microphones are
out of phase. If correctly wired, the level should be
slightly higher when both mics are switched on

Figure 5.3: Test for microphone phase

small spacing between the mics isn't causing any problems. If sound is arriving at one microphone slightly before it arrives at the other then some phase cancellation will occur, but only at high frequencies where the wavelength is short – at lower frequencies these effects are negligible. Figure 5.3 shows how this test might be arranged.

Phantom Powering

All capacitor microphones and electret microphones require power in order to operate, and although electret mics will work from a low enough voltage and require little enough current to make battery powering practical, true capacitor microphones normally utilise an external source of power because of the higher voltages needed. The standard power supply voltage for microphones is 48v, though some models are designed to operate over a range of voltages, from 9v to 48v, allowing them to be operated from different power supplies.

The most popular way of powering a capacitor mic is to use a phantom power supply. This ingenious system supplies the necessary DC power along the signal leads in the microphone cable, which means that no extra wiring is necessary. However, the mic wiring must be balanced to accommodate this system.

A phantom power supply may be a stand-alone unit that is connected in series with the microphone cable at some convenient point or, as is more often the case, it is generated within the mixing console or microphone pre-amplifier and fed via the microphone input sockets. Less expensive consoles allow the phantom powering to be switched on or off globally, while more intricate designs allow individual switching for each mic input.

Connecting a dynamic mic to a phantom powered input will cause no problems as long as the mic is wired for balanced operation, so that the same voltage appears

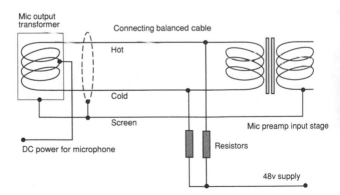

Figure 5.4: Phantom powering schematic

at both ends of the voice coil and no current flows; but if the mic is wired for unbalanced use then the phantom powering voltage will be present across the coil and a potentially damaging current will flow. Even if the mic is wired for balanced use, it's good practice not to plug it in with the phantom power on just in case one pin makes contact before the other, causing current to momentarily flow through the coil. This is unlikely to cause any damage but it's better to be safe than sorry! Figure 5.4 shows the schematic of a typical phantom powering system.

6 SPECIAL TYPES OF MICROPHONE

So far we have looked at fairly conventional types of microphone, which are typically stand mounted or hand held and are located fairly close to the sound being recorded. Problems can occur if these mics are mounted at a greater distance, as sound reflecting from walls or other solid objects arrives at the mic slightly later than the direct sound, causing partial cancellation of some frequencies. This gives rise to peaks and troughs in the output frequency, and an often-used term for this effects is 'comb filtering', a more severe version of something deliberately created in phaser and flanger units and which can result in a distant and highly unnatural sound.

Boundary Microphones

One microphone which neatly avoids this problem is the PZM (Pressure Zone Microphone), initially developed by Crown. Other companies produce similar designs, but these are generically termed boundary-effect microphones to avoid copyright problems.

The boundary-effect microphone, is designed to be

used at a boundary such as a wall or floor, but models with integral boundary plates are also available which can be positioned more freely. It should be noted, however, that the plate's size affects the low-frequency response of the mic, so the mic should still be mounted on a plate at least one metre square if the whole audio bandwidth is to be faithfully reproduced. At frequencies below which the boundary plate is effective, the effect ceases to work and a drop in low-frequency response is evident. The response usually shelves off by 6dB, when the wavelength of the sound is longer than six times the length of the side of the plate (assuming a square plate).

The Boundary Effect

The principle of the effect relies on the fact that, at a solid boundary, there can be no air movement because the mass (and consequentially the inertia) of the boundary prevents it. At such boundaries reflected sound manifests itself instead as a change in air pressure, and a pressure-type microphone capsule located at the boundary can accurately respond to these changes.

Some designs use a pressure capsule recessed so as to be flush with the plate, while others rely on a capsule suspended at a small distance above the plate and facing towards it. It is this latter approach that is used in the popular (but now sadly discontinued) Tandy/Radio Shack PZM, in which the capsule senses pressure changes in the gap between the capsule and the boundary. As long as the capsule is within a tenth of an inch of the boundary plate, the direct and reflected sound will remain in phase up to 20kHz.

The advantage of this design is that direct and reflected sounds arrive at the capsule simultaneously, which gives an inherently consistent frequency response over the whole hemisphere of the microphone's pickup pattern. Furthermore, the direct and reflected sounds arrive in phase, and so combine to double the output of the mic.

Polar Pattern

You can think of the boundary mic as being like an omni mic with half of its pattern cut off by the boundary. Figure 6.1 shows how a typical boundary-effect microphone might be constructed and how direct and reflected sound would arrive at the capsule.

Directional boundary mics are also available which incorporate a cardioid capsule, usually pointing along the plate, where the advantages of in-phase direct and reflected sound are maintained.

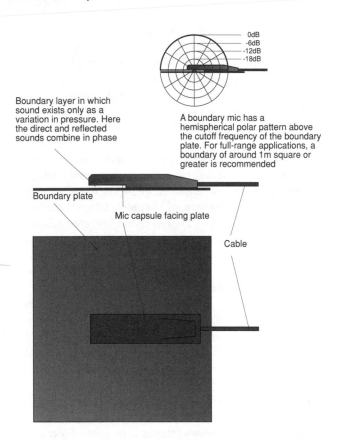

Boundary layer in which sound exists only as a variation in pressure. Here the direct and reflected sounds combine in phase

A boundary mic has a hemispherical polar pattern above the cutoff frequency of the boundary plate. For full-range applications, a boundary of around 1m square or greater is recommended

Boundary plate

Mic capsule facing plate

Cable

Figure 6.1: Boundary-effect microphone

7 STEREO MIC TECHNIQUES

Most of what we hear on pop records is not true stereo but a collection of separately recorded mono sounds panned around different positions in the left/right soundfield. What's more, these sounds are frequently enhanced by digital reverberation using synthesised stereo outputs. However, there's a world of difference between true stereo and what I think of as panned mono.

The underlying theory of stereo is rooted in the fact that we have two ears. The reason we have two rather than just one is so that we can tell from which direction sounds are coming. The manner in which we differentiate between those sounds arriving from the front or from behind is a complex and involved process, and one in which the shape of the outer ear is just one factor of many, but the most important parameters covering left/right positioning are fairly well understood.

Direction

If we hear a sound that is directly in front of us, then as long as there are no strong reflections from nearby objects it is reasonable to assume that each ear will receive the same sound at the same time, because the human head is symmetrical. However, if we hear a sound coming from the right, that sound will reach the right ear a fraction of a second before it reaches the left. Furthermore, when the sound does reach the left ear, it will be slightly quieter due to the masking effect of the head and its frequency content will be altered (low frequencies tend to pass around the head, whereas high frequencies will be absorbed or shadowed).

Precedence

Our brains are equipped to analyse this level, phase and spectral information, and this is what enables us to estimate the direction from which a sound is coming without any conscious effort on our behalf. Of course, the introduction of reflecting surfaces can confuse the listener, but there is yet another effect of nature on our side: the precedence effect.

If two sounds of equal intensity are separated by more than around 15ms, but are still close enough to sound like a single event rather than a sound followed by an echo, the brain locks onto the first sound to arrive and works out the direction of the source from that. In a large, reverberant room, the original sound is often far enough ahead of the first reflection to allow the precedence effect to work, but in a small, highly reflective room the result may be more confusing.

Dummy Head

It might seem from the above description of human hearing that the only way to record a concert in true stereo would be to sit in the audience with a small microphone jammed in each ear. Surely this would capture exactly the sound the listener would normally hear? Bizarre though it may seem, work has actually been done on this, utilising tiny microphone capsules fitted within the ear, and when the sound is replayed over headphones it appears to retain many of the stereo cues of the original performance. A far less uncomfortable approach, however, is to build a dummy head and build the microphones into that.

Promising though this line of research appears at first glance, there is one serious limitation: if the sound is picked up from within the ear then to accurately play back a recording the sound must be generated in the ear by tiny speakers positioned in exactly the same positions as the microphones. Although this isn't practical, the use of headphones is close enough to the theoretical ideal to yield excellent results, though front and rear sounds still appear to be confused. Several records have been released that used dummy-head recording, and it is worth listening to some of these through really good headphones just to hear what can be achieved.

Loudspeaker Reproduction

Unfortunately, the technique doesn't work nearly as well when the sound is replayed via loudspeakers. This is due to the fact that, when we hear anything through stereo loudspeakers, some of the sound from the left speaker reaches the right ear, and vice versa. To create a convincing stereo image with loudspeakers, different microphone techniques are necessary.

To this day, we still don't have an entirely satisfactory system for capturing and reproducing a soundfield via a pair of stereo loudspeakers. After all, when we attend a concert, sound arrives at us from a wide variety of directions – left/right, up/down, front/back and every conceivable angle in between. When we listen to speakers, all of the sound comes from just two small boxes, perhaps with a little room reverberation thrown in.

Spatial Recording

One can speculate on how a true spatial recording might be made in the future, perhaps by covering the surface of a sphere with microphones and then replaying the recording via a similar sphere, the inside of which was entirely covered with loudspeakers. The listener would occupy the sphere's centre. The real problem lies in intercepting all of the sound arriving at the listener and then recreating that same soundfield at the same points in space when we come to replay the recording. This is rather like the audio equivalent of a hologram, but to date the technology doesn't exist to pursue this avenue, at least not in any practical way.

Fortunately, our ears are fairly forgiving organs, and there are a variety of stereo recording techniques that produce a reasonably convincing illusion of space without going to all of this trouble.

Coincident Or XY Pair

Probably the oldest (and still one of the most popular) methods of stereo miking is the coincident or XY pair. This comprises two high-quality cardioid or figure-of-eight mics of similar characteristics mounted at around 90º to each other and with their capsules as close to each other as is practically possible. Because the microphones are directional, one will pick up sound mainly from one side of the soundstage while the other will pick up sounds from the opposite side. If figure-of-eight mics are used then it's also possible to capture the left and right sounds from the rear of the room, including audience noise and room reverberation.

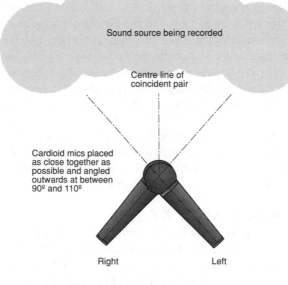

Sound source being recorded

Centre line of
coincident pair

Cardioid mics placed
as close together as
possible and angled
outwards at between
90º and 110º

Right   Left

Figure 7.1: Coincident cardioid mic array

This arrangement provides the necessary change in
level from left to right but makes no serious attempt
to simulate the masking effect of the human head or
the phase effects caused by the distance between one
ear and the other. However, because cardioids often

have a poor high-frequency off-axis response, the high-frequency attenuation caused by the presence of the human head may be simulated to some extent, purely by accident! Figure 7.1 shows a basic coincident pair set up relative to a sound source.

Phase Integrity

One of the most important advantages provided by the coincident or XY system is that the lack of spacing avoids any phase problems that might compromise mono compatibility. There are drawbacks, however: firstly, the lack of the phase information dilutes the stereo effect; and secondly, because cardioids and figure-of-eight mics tend to have their most accurate frequency response on or near their axes, the important central area of the soundstage may end up sounding less clear than the edges. The degree to which this occurs depends on the off-axis characteristics of the particular microphones used. Despite these shortcomings, however, the coincident pair is widely used and is capable of excellent and predictable results. What's more, the negligible spacing between the two mics means that mono compatibility is good.

When using figure-of-eight microphones rather than cardioids, it must be borne in mind that there is no way of discriminating between front and rear sounds, so the replayed recording will always present all of the information in front of the listener.

The angle subtended by a coincident pair is about 90°, and sounds occurring outside this angle will be represented inaccurately in the stereo recording. The angle can be widened slightly by increasing the mic angle to 110°, but any further widening is likely to be at the expense of the centre-stage sound quality.

Middle And Side (MS)

There is another coincident microphone technique which produces a more stable centre image than the XY pair and which is completely mono compatible. Figure 7.2 shows the arrangement, which again uses two coincident microphones, this time one figure-of-eight and one omni. In situations in which sound from behind the mic array should be excluded the omni may be exchanged for a cardioid.

This system uses one microphone, the 'middle' of which captures the centre-stage sound in mono. The figure-of-eight microphone is positioned so that it 'listens' left and right, which is why this is called the side mic. Essentially, the middle mic provides an accurate mono image, while the side mic provides the information that, when properly decoded, describes

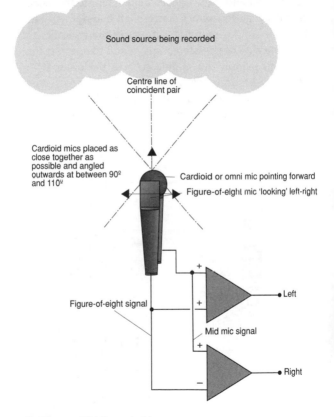

Sound source being recorded

Centre line of
coincident pair

Cardioid mics placed as
close together as
possible and angled
outwards at between 90º
and 110º

Cardioid or omni mic pointing forward

Figure-of-eight mic 'looking' left-right

Figure-of-eight signal

Mid mic signal

● Left

● Right

**Figure 7.2: Middle-and-side
mic system**

how the left and right signals are different to the mono centre signal. To extract discrete left and right signals it's necessary to sum the outputs of the two mics to give one side of the stereo image and to subtract the side signal from the middle to give the other side. This may be done by using a specially constructed sum and difference box, or it may be done by using a mixing console that has phase reversal buttons, as shown in Figure 7.3.

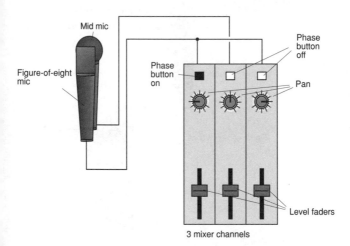

Figure 7.3: Decoding MS outputs using a mixing console

Note that three mixer channels are needed to do the decoding: the middle mic is panned dead centre while the side mic is split to feed two channels, one panned left and the other panned right. On one of these channels the phase button is pressed, and this pair of channels, one in phase and the other out of phase, gives us the necessary sum and difference signal.

Variable Width

The stereo width can be modified by varying the level of the side signal channels. With the side signals turned right down we are left with mono, and if the side signals are turned up higher than the centre signal we end up with an artificially widened stereo image. When using a mixer care must be taken to set the two side channels to the same level, and this may be done by panning both to the centre (with the mid signal switched off) and adjusting the levels so that the two signals cancel each other out. Once this has been done, the channels may be panned back to their original left and right positions.

For location recording, it's quite in order to record the outputs from the two microphones directly onto two tracks of a stereo tape recorder, and these may be decoded later when mixing or transferring to another stereo machine.

The MS pair has the advantage of being always mono compatible – the in- and out-of-phase side signals cancel completely when summed to mono, and the centre-stage signal is captured with greatest accuracy because it is directly on the axis of the mid mic. Despite the minor inconvenience involved in decoding the output signals to give the left and right stereo signals, the technique can be very rewarding and is highly recommended.

However, because this is still a coincident technique, there is no spacing to yield phase information. Furthermore, the figure-of-eight mic picks up a lot of its information off axis, so we have a situation which is the reverse to that encountered with the XY pair: the centre-stage sound is accurate, but the edges of the soundstage may be less well defined.

Spaced XY

So far I have looked at systems that create a level difference between left and right sounds, but no attention has been paid to the time delay caused by the distance between the ears. We can space the microphones of our coincident pair by a few inches to create an additional sense of space, but the phase cancellations that occur should the signal be summed to mono cause comb filtering effects, which may significantly colour the sound. If there is no need to reproduce the recording to a high standard in mono then the technique is perfectly acceptable. In fact, this system is favoured by some European broadcast institutions, where it is known as the ORTFE (Office de Radiodiffusion Télévision Français) system. To be ORTFE compatible the microphones must subtend an angle of 100º and be precisely seven inches or 17cm apart. A variation on this setup is used by the Dutch Broadcast Organisation – their NOS pair subtends 90º, with a 12in or 30cm spacing.

Spaced Omnis (AB)

If, on the other hand, we use omni mics and space them by several feet rather than inches, the comb-filtering effects become less pronounced, except at low frequencies, where they still persist. This is a popular stereo technique known as the AB pair, and is used for recordings destined to be replayed over loudspeakers, although positioning is important if the soundstage isn't to suffer from a 'hole' in the middle caused by too great a mic spacing. If such a hole is evident then a third omni mic can be added centre stage, but then additional care has to be taken to avoid problems with phasing. If such problems persist then all you can do is change the mic positioning and spacing so that the effect is minimised. Try to keep the inter-mic spacing

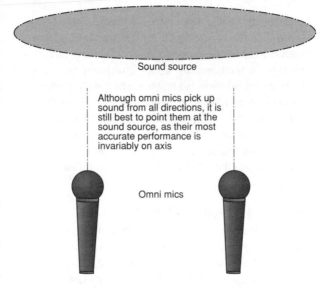

Sound source

Although omni mics pick up sound from all directions, it is still best to point them at the sound source, as their most accurate performance is invariably on axis

Omni mics

Figure 7.4: Spaced omnis set up to record a small ensemble

as large as possible relative to the distance between microphone and sound source. Figure 7.4 shows a spaced omni setup which could be used for a small musical ensemble.

The distance of the microphones from the performers must be carefully adjusted so that the desired balance of direct and reverberant sound is picked up. If the sound is too ambient then the mics need to be brought closer to the performers. Conversely, if the sound is too dry, the microphones should be moved further away. Don't move the mics too far in one go, however, because a relatively small change in distance can significantly change the sound. If you can arrange to monitor a rehearsal using good headphones while you fine tune the mic positions then this will save a lot of trial and error.

Spot Microphones

With very large ensembles it is sometimes impossible to achieve the desired balance by using just a stereo pair of mics, so additional spot mics are often brought into use. These are positioned close to those performers needing reinforcement and then added to the main stereo mix after being panned to the appropriate position. The technique for positioning these spot mics is exactly the same as it would be if you were miking an individual or small group of players, and either cardioids or omnis are applicable.

A potential problem in employing spot mics is that the sound arrives at the spot mics several milliseconds before reaching the main stereo mic array (remember that sound travels at just over one foot per millisecond).

This can confuse the imaging and cause phase cancellation effects. If the problem is serious enough to warrant attention, a high-quality digital delay line can be used to delay the spot mics and bring them back into line with the stereo mics. A stereo delay patched into a pair of mixer subgroups would be one practical way to achieve this, and around a millisecond of delay should be added for each foot between the spot mics and the main stereo array.

Whenever spot mics are used they should be added to the stereo mix in as small a proportion as possible so that the main stereo mics still provide the majority of the signal. This will prevent the stereo image from becoming too confused, which could happen if several high-level spot mics were competing with the stereo pair.

Stereo With Boundary Mics

Boundary mics are capable of producing excellent stereo recordings and, like conventional mics, there are a number of different ways of obtaining essentially the same result. The simplest method is the equivalent of spaced omnis, and two boundary mics placed four feet or so apart on a low table top make an effective and easy method of recording a solo player. Alternatively, the two boundary mics may be fixed to a wall and separated by the necessary distance to form an equilateral triangle, with the performer at the apex. Both arrangements are shown in Figures 7.5a and 7.5b. It is also possible to mount the microphones directly on the floor, but this leaves them open to the danger of picking up low-frequency vibrations and thumps from the floor itself.

The boundary mic equivalent of a coincident pair is the arrangement whereby two mics are fixed to either side of a wooden or perspex sheet around four feet square. The mics should ideally be positioned a few inches away from the exact centre of the board to prevent edge-diffraction effects, and the board should be suspended edgeways to the performers, in a position similar to that which would be used for coincident cardioids. Again the exact distance has to be determined empirically so that the desired balance of direct and reverberant sound is picked up.

Rear Pickup

This setup will pick up sounds from the front and rear of the room with equal efficiency, so if it is desired to reject some of the rear sound – for example, audience noise at a live concert – then two sheets of material may be employed, hinged at the front edge. By increasing the angle between the two sheets, the mic array can be made less sensitive to sound coming from

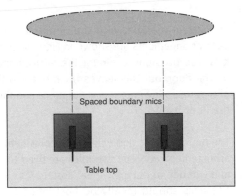

Spaced boundary mics

Table top

Figure 7.5a: Spaced boundary mics on a table top

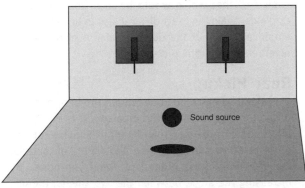

Boundary mics fixed to a wall

Sound source

Figure 7.5b: Spaced boundary mics fixed to a wall

Boundary mics fixed at either side of a board positioned edge on to the sound source

Figure 7.6a: Boundary mics fixed to either side of a sheet for stereo pickup

Direction of sound source

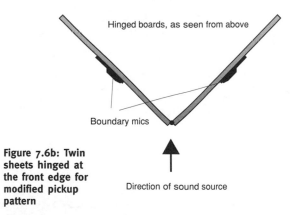

Hinged boards, as seen from above

Boundary mics

Figure 7.6b: Twin sheets hinged at the front edge for modified pickup pattern

Direction of sound source

behind at the expense of narrowing the image of sounds arriving from the front. Once again, it is up to the user to choose a suitable compromise, and both methods are illustrated in Figures 7.6a and 7.6b.

8 RECORDING VOCALS

Most studio professionals will use capacitor mics for vocal use because of their high sensitivity and wide frequency response. Dynamic microphones tend to perform poorly above 16kHz or thereabouts and are less sensitive than capacitors, which could lead to problems with noise if the vocalist has a quiet voice. On the other hand, some rock vocalists prefer to use their live dynamic mic in the studio because it gives them a hard-hitting, powerful sound and their voices are usually loud enough not to require a highly sensitive mic. Here the choice is made for artistic reasons rather than to capture the most accurate rendition of the performer's voice.

Frequency Response

One seldom-aired point is that the wide frequency response of capacitor microphones can emphasise sibilance in a performer's voice (a whistling sound accompanying 'S' and 'T' sounds), and modern bright-sounding digital reverb units tend to further exacerbate the situation. The traditional cure for this problem is to use an electronic de-esser to attenuate the sibilant sounds. However, a more pragmatic approach might be to use a suitable dynamic microphone on performers known to suffer from excessive sibilance, as the limited frequency response will help to hide the problem. Moving the position of the mic so that the performer sings over rather than directly into the microphone can also help avoid sibilance.

Polar Pattern

In the context of a live performance, I've indicated that cardioid pattern mics are the preferred choice because of their ability to reject off-axis sounds, thus minimising the spill from other instruments or performers. They are also less prone to acoustic feedback. In the studio, vocals tend to be recorded as separate overdubs, during which the singer monitors the existing instrumental backing via headphones. Because of this, the need to use a cardioid pattern mic is not as great, but unless the studio is acoustically fairly dead then it may be a good idea to use a directional mic anyway to minimise the effect of the room acoustic on the recorded sound. If the acoustic is compatible with the use of an omni pattern mic then the result will probably sound more natural because, as a general rule, omni-pattern mics produce a more accurate result than an equivalent cardioid. Also, because omni mics have the same nominal response in all directions, any room ambience which is captured will sound more accurate than it would if a cardioid were used. The off-axis response of a cardioid mic can exaggerate the boxy sound of a room.

No single microphone is ideal for all vocalists. If a singer has a bright voice then a mic with a presence peak may tend to make the overall sound harsh, whereas the same mic used on a person who has an indistinct or soft voice could yield more favourable results. I would recommend recording vocals with little or no EQ, and then, if the sound isn't right, try a different mic or change its position slightly before resorting to equalising the sound.

Handling Noise

To minimise handling noise, microphones should be mounted on a solid boom stand, and a shock-mount cradle should be employed if possible. If your mixing console has a low-frequency hi-pass filter, this can be used to reduce the effect of any low-frequency vibrations transmitted from the floor via the stand, but it's better to avoid them in the first place.

Pop Shields

Many people experience problems with popping on the plosive 'P' and 'B' sounds, and experience has shown that simple foam wind shields are generally ineffective. The popping is caused by blasts of air from the performer's mouth slamming into the microphone's diaphragm, giving rise to a high-level, low-frequency output signal, which manifests itself as a loud, breathy pop or thump. Fortunately, correcting this problem is simple.

Commercial pop filters simply comprise a fine plastic or metal gauze stretched over a circular frame somewhere between four and six inches in diameter and positioned between the performer and the microphone. Normal sound (which consists of vibrations within the air) is not affected, but blasts of moving air are intercepted and their energy is dissipated as turbulence as the air forces itself through the small holes in the mesh. A commercial pop shield may seem expensive, but it is possible to improvise your own at a fraction of the cost which will work just as well.

The important part of the pop filter is the grille or mesh, and a piece of nylon stocking works perfectly. This may be stretched over a wire frame (many engineers use wire coat-hangers) and then fixed around six inches in front of the microphone. Figure 8.1 shows a typical setup. An alternative ready-made pop filter can be bought in the form of the splash guard of a frying pan, and this simply consists of a fine metal mesh fixed to a wire

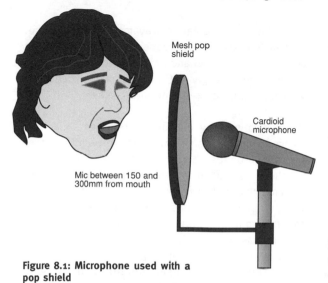

Mesh pop
shield

Cardioid
microphone

Mic between 150 and
300mm from mouth

**Figure 8.1: Microphone used with a
pop shield**

hoop with a handle. The provision of a handle makes
fixing the filter to a mic stand a simple task, and the
overall result is a little tidier than the stocking approach.

Positioning

There is generally no need for a singer to stand really
close to a microphone in the manner of live
performances, though those performers who have
mastered the art of using the proximity effect of the

microphone in a creative way may wish to do so. Working too close to a microphone will cause significant changes in both level and timbre as the singer moves his head, so as a general rule only very experienced vocalists should be recorded in this way.

Normally, a working distance of between six and 18 inches is ideal, with the microphone pointing either just above or just below the singer's mouth. This usually gives a little more immunity to popping and sibilance than pointing the mic straight at the mouth does, though the singer should take care not to move too far from the main axis of the microphone's response, as most cardioid mics only have an accurate frequency response over an angle of plus or minus 30° or so. Outside this range, therefore, the top-end response usually falls off, with the consequence that off-axis sounds are reproduced sounding duller than they should.

The best position for a stand-mounted microphone is in front of and just above the performer's mouth. If the stand will permit the mic to hang, rather than supporting it from underneath, the vocalist will find it easier to arrange lyric sheets so that they can be consulted during the performance without changing the head position.

Acoustics

Most studio vocals are recorded in a relatively dead environment so that the desired reverb can be added artificially during the mix. If the acoustic environment in which you are working is adversely affecting the recorded sound, it's possible to achieve excellent results by applying deadening material to one corner of a room and then getting the performer to stand in the corner singing towards the centre of the room. Any sounds reflected from the rest of the room and bounced back into the corner will be absorbed by the deadening material before they can be bounced back into the microphone (as long as a cardioid microphone is used). Acoustic foam tiles are very efficient at the frequencies normally found in speech, and can be fixed either directly to the wall or onto the surface of a hinged screen. If such a screen is made with a reflective flipside (of hardboard or varnished wood), then it may be used to deaden an environment by using the foam tiles or flipped over so that the wooden side is face-up to increase the room's liveness. Figure 8.2 over the page illustrates how such a corner that has been acoustically treated in this way might be used in practice. However, it is inadvisable to move the arrangement too far into the corner of the room, as the boundary effect of the walls can introduce an unwanted low-frequency boost.

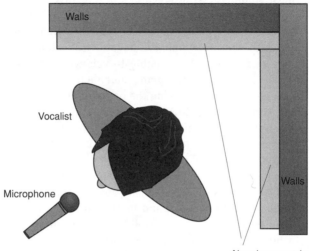

Walls

Vocalist

Microphone

Absorbers may be either foam tiles or improvised from heavy curtains or bedding

Walls

The absorbers behind the singer prevent sounds reflecting back from the walls into the microphone

Figure 8.2: Use of acoustic treatment in a corner for recording vocals

Multiple Vocalists

If it is necessary to record more than one vocalist simultaneously, it is advisable to position the mics so that any spill between them is kept at a minimum (unless you are using a single stereo pair). The distance between mics should be at least three times the distance between the microphone and the vocalist in order to avoid phase cancellation effects, and some form of acoustic screening is also advisable. This three-to-one rule applies whenever multiple microphones are used in close proximity.

If only two performers are being recorded at the same time, it is quite acceptable for a figure-of-eight mic to be used, with a singer on each side. This setup avoids any phase-cancellation problems but also precludes the rebalancing of levels at the mixing stage.

With larger groups of singers, such as choirs or ensembles, it may be desirable to record them in stereo, and suitable techniques for this are discussed in the section on stereo miking. As a rule, one stereo pair should be used for each section of the chorus, with the pairs of microphone mounted above the heights of the performers' heads and several feet in front of the front row. Figure 8.3 shows how several stereo pairs might be used to mic up a large choral group.

Musical ensemble comprising several sections

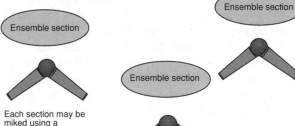

Each section may be
miked using a
separate coincident or
spaced pair. Individual
soloists may be
covered using a single
microphone

An additional stereo pair may
be used to capture the overall
sound. Some engineers prefer
to delay the close mics so as to
get the signal in phase with the
more distant mics

Figure 8.3: Multiple stereo pairs used to record a large choral group

Processing

Because most vocalists find it difficult to control their dynamics with any accuracy, some compression may be needed during the recording to ensure a healthy signal level on the tape (or digital recording system) at all times while arresting any peaks that might cause overload distortion. Soft-knee compressors are ideal for use with analogue recorders, though digital recorders may also benefit from being used with a separate limiter to avoid brief peaks that the compressor might allow through. As a starting point, try setting the compressor to give a gain reduction of eight or so during the louder sections, although a performer with a very wide dynamic range may need a ratio type controller set to a ratio of 4:1 or even higher. The same degree of gain reduction should be attempted, and, as with all signal processing, if you can get away with using less then this is preferable, as you'll almost certainly end up with a more natural sound. Always under- rather than over-compress when recording, as you can always add more while mixing, but reversing the effects of excessive compression is far more difficult.

Headphones

It is also worth writing a few lines on the subject of monitoring. Fully enclosed headphones are the preferred choice, as they are less prone to spillage problems, but some vocalists find that they have trouble singing in

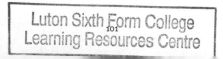

tune when using this method. This is why you often see videos of recording sessions in which the singer has one phone on and the other off.

Semi-enclosed phones through which you can actually hear feel more comfortable to work with but they leak more sound, so you could end up with a small but audible level of the backing track on the vocal track. This is acceptable as long as you don't later decide to use part of the take whilst unaccompanied, because the spill will become obvious when there is no music to hide it.

The same is true if any sort of click track is being used, and the mid-range beeps provided by some sequencers tend to spill quite badly, even with enclosed headphones.

Alternative Monitoring

For those vocalists who simply cannot work with phones at all, there is one method of working with speakers that gives surprisingly little spill. If the backing track is switched to mono and the phase of one of the monitor speakers reversed by swapping the red and black terminal connections on the rear of the speaker cabinet, then a microphone positioned equidistant from both speakers will pick up an equal amount of in-phase and out-of-phase sound. Theoretically, this should cancel out, leaving no signal picked up by the mic. However, nothing is perfect, and so some signal is always audible, but by carefully positioning the mic and by balancing the levels fed to the speakers the amount of spill can be reduced to a bare minimum. Because the vocalist is listening to the speakers with both ears, the out-of-phase effect doesn't cause cancellation and the monitored sound is perceived as being quite normal. Figure 8.4 shows how this arrangement could be set up in practice.

Spoken Word

Spoken-word recording is more critical of room acoustics than singing is because, in the latter case, clarity of diction is not quite as important, and there are likely to be other instruments playing which will hide the more subtle room effects and cover extraneous noises. Ideally, a very dead acoustic is necessary, and again this can be arranged locally by means of acoustic tiles, a portable sound booth or by hanging blankets or rugs around the recording area.

A high-quality capacitor microphone is the ideal choice, and a cardioid model will exclude more room ambience than an omni would. If some degree of reverberation is necessary, this may be added artificially after the recording has been made. Reverb should be used very

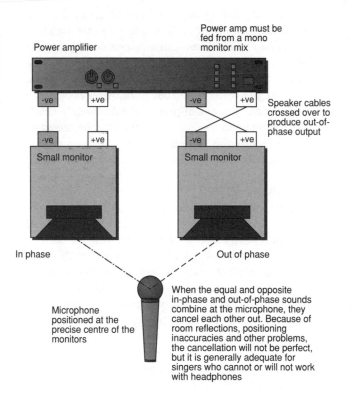

Figure 8.4: Using out-of-phase monitors to avoid spill

sparingly unless a special effect is sought and, dedicated ambience programs often work better than more obvious reverb treatments.

Early Reflections

It is particularly important to minimise early reflections from nearby objects such as tables or walls, and it is common for specially absorbent or acoustically transparent table tops to be employed. A table top could be made absorbent by placing a two-inch layer of foam rubber on the surface and transparent by using a perforated metal sheet or wire mesh. A table is generally necessary to support scripts or notes, and care should be taken to ensure that these don't rustle during recording.

If a solid table must be used, an improvement in early reflections can be achieved by using a boundary mic flush with the table top. This arrangement ensures that reflections from the table will either miss the microphone altogether or arrive in phase with the direct sound, which avoids comb-filtering effects and their inevitable coloration.

Multiple Voices

Where it is necessary to record several speakers at once, there are a choice of approaches that can be used. In conferences, where the main objective is to provide a record of what has been said, a single omni hanging above the table where the speakers are seated will often suffice. A location about two feet above head height is normally adequate. This method has the advantage of simplicity but allows no balancing between the voices, and the result is only monophonic.

For the recording of two voices, you can use either a figure-of-eight mic with a speaker at either end of it or two cardioids, ensuring that there is adequate space between them to maintain separation and prevent phase cancellation from colouring the sound.

The figure-of-eight option is the easiest and has no phase problems, but the result is again mono. Also, the only way to balance the levels is to experiment with the position of the speakers during a test run. If their voices are significantly different in level, then separate microphones will yield a better result.

Separate microphones have the advantage that they can be recorded onto different tape tracks to create a stereo spread, and the levels can easily be adjusted individually. Either technique is suitable for commentary and basic radio play or jingle work.

9 ACOUSTIC GUITARS

Acoustic guitars and similar fretted instruments often have a particularly limited sound output, and so therefore you should use a sensitive microphone if you want to avoid problems picking up undue amounts of noise. The sound of the acoustic guitar also contains a lot of high-frequency detail, and so using a capacitor or good back-electret microphone should be the obvious first choice, and you should preferably use one that has a reasonably flat frequency response over the entire audio spectrum.

When miking an acoustic guitar in a live setting, it seems to be the obvious choice to put the microphone very close to the sound-hole, as this is the part of the instrument which produces the largest signal. Unfortunately, however, the tone picked up here is likely to be boomy and unpleasant. Other important components of the sound come from elsewhere on the instrument, such as the top of the body and the strings, and it's only when they all combine that the true guitar tone is created.

Most dynamic microphones are not sufficiently sensitive to do justice to the acoustic guitar, and their insufficient top-end response can sometimes result in a somewhat lifeless sound. If you must use a dynamic microphone, however, it's important to use a good-quality model and resist the temptation to position it too close in order to get a healthy signal as this will seriously compromise the tone.

Vibrations

Like most acoustic instruments, the various parts of the guitar vibrate in different ways, which means that different parts of the instrument generate different sounds: the volume of air inside the body produces one sound, the vibration of the table produces another and yet further sounds are produced directly from the strings and from the vibration of the neck and headstock. What the audience hears is a blend of all of these sounds, along with reflections from the floor and walls of the room itself, so it is evident that placing a microphone close to a single spot on the instrument's surface won't give a representative sound. This is why contact mics and bridge transducers seldom produce a sound that matches up to the natural tone of the instrument, but judicious positioning, combined with careful EQ circuitry, can produce a close approximation. Though serious recordings are rarely made using transducers alone, it is fairly common practice to combine the output from a transducer with a mic.

Mic Patterns

Either omnidirectional or cardioid mics may be used, although omnis are preferable only when the room acoustics are particularly good. Strictly speaking a flat frequency response is desirable, but a microphone with a subtle presence peak might enhance the top end, allowing it to cut through a mix. The tone acquired by choosing and positioning a suitable mic always sounds better than a tone acquired by using EQ. If two guitars are being recorded together, then it may be worth trying to use omnis rather than cardioids. Then the degree of spill will be higher, but, due to the superior off-axis response of omnis, it will at least be of a high quality. In such a situation, use the three-to-one rule for mic positioning: if the mics are placed two feet from the guitars then make sure that they are at least six feet apart. If the microphones are positioned any closer than this then the degree of spill may be enough to cause phase-cancelling effects that will end up colouring the sound, making it distant and unnatural.

Environment

In order to produce the best tone from the instrument, it should be played in a sympathetic room – one in which the acoustics are fairly live. If no such room is available, a localised live environment can be created by covering the floor with hardboard, with the shiny side facing up, and by placing reflective screens around the playing area. I've even heard of corrugated aluminium sheeting being used to create a reflective environment, the main criterion being that the instrument actually sounds good in the room in which it is played. Don't worry if the acoustic isn't as live as you'd like as some artificial reverb can be added without compromising the quality of the sound.

Mic Placement

There are any number of possible methods of arranging mics that produce excellent results, but a mic positioned around two feet from the instrument and pointing somewhere between where the neck joins the body and the sound-hole is a good starting point. If the sound is too boomy then the mic can be positioned further away from the sound-hole or moved upwards so that it is 'looking down' on the instrument; and conversely, if the sound lacks body, it can be moved closer in.

Try moving your head around the playing position as the performer warms up and listen out for any 'sweet spots'. When you've identified these, try putting the microphone there and listen to the quality of the sound at that spot. If it's not quite right, you can always move it to a better position.

A quick method of finding out the exact location of these sweet spots is to monitor the output from the mic using good-quality closed headphones and then move the mic manually as the guitarist runs through the number. In this way you can immediately hear the effect that changing the microphone's position has on the final sound.

Stereo

If the guitar is the main instrument in a track then try recording it in stereo as this will give a much more detailed sound. To achieve this, place a stereo pair of microphones two or three feet in front of the instrument and move the mic positions slightly if the sound is excessively thin or boomy. Remember that pointing a microphone towards the sound-hole will give a more boomy sound, whereas moving the mic along the neck and away from the body will give a thinner sound. Figure 9.1 shows some possible stereo mic positions. In practice you can use any of the stereo miking methods discussed in the sections on stereo miking.

It is also quite acceptable (and indeed probably more common) to use two mics which can be panned to

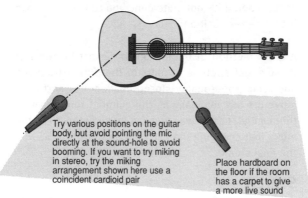

Try various positions on the guitar body, but avoid pointing the mic directly at the sound-hole to avoid booming. If you want to try miking in stereo, try the miking arrangement shown here use a coincident cardioid pair

Place hardboard on the floor if the room has a carpet to give a more live sound

Figure 9.1: Stereo-miked acoustic guitar

produce a stereo image, even though it may not be possible to position them to capture a true stereo picture. For example, you could have one close mic panned to one side and an ambience mic positioned further away and panned to the other. Unless you are looking for an accurate stereo picture, any setup that sounds good can be considered artistically valid. Similarly, panning the miked signal to one side and the output from a transducer to the other often produces worthwhile results.

Practical Arrangements

One arrangement that works particularly well is positioning one mic in the usual end-of-neck or looking down position to capture the main part of the sound, and then to position another a foot or so away from the neck pointing towards the headstock. This gives a very articulate, detailed sound that works equally well with picked, rhythm or classical playing. The mics may be cardioids, omnis or even one of each, and they may be balanced during the mix to create the desired tone. Figure 9.2 shows how this arrangement can be set up.

Interesting results can be achieved by positoning one mic to cover the neck and headstock of the guitar

Figure 9.2: Asymmetrical mics used to create a stereo effect

If you have a pair of boundary mics then you can obtain a very natural sound by placing these on a table top a couple of feet in front of the performer with the mics around three feet apart. This arrangement is also quite good for making quick demos of guitarist/vocalists. Alternatively, the boundary mics could be set up as a conventional stereo pair, as outlined in Chapter 7, 'Stereo Mic Techniques'.

Preparation

As with any instrument, the resulting sound quality depends on how good the guitar sounds in the first place. Any intonation problems or buzzes should be addressed before the session starts, and with steel-strung instruments then using a relatively new set of strings is a good idea, particularly if a bright, lively sound is sought.

The performer should play while seated, if possible, to prevent undue movement of the guitar relative to the microphones, and it should be ensured before recording begins that the seat doesn't squeak or creak. Also care should be taken to exclude other sources of noise, such as spill from headphones, particularly if click tracks are being used. Some finger squeak is normal, but if excessive then a little talcum powder on the player's hands may help.

External Noise

Other sources of noise include the ticking from clockwork or even quartz watches, rustling clothing and excessively loud breathing. Some breathing is obviously a natural part of the performance, but some players tend to accompany their music by grunting and snorting. If a click track has to be used then pick a sound that isn't too piercing, and if possible rig it via a compressor set up in 'duck' mode so that the click track gets louder when the guitar is played louder and gets quieter when the playing is quiet. In this way the performer will always be able to hear the click, but it will be turned down during quiet sections and pauses, when it's most likely to be noticed.

Equalisation

In an ideal world equalisation would not be necessary, but in the studio, where reality is not always the main objective, some EQ may prove essential. Even so, try not to use more than the barest amount.

Booming is a large problem with certain acoustic guitars, and you can find out the frequency at which this occurs by turning up the boost on your lower-mid or bass equaliser and then sweeping through the frequency range until the booming is picked out. Once you've found this frequency, which will probably lie between 100Hz and 200Hz, you can apply a little cut to lessen the effect. Hopefully a decibel or two of cut will be sufficient.

If the sound needs a little more top-end sparkle then you could use your shelving 10kHz equaliser to add a decibel or so of top boost. If this doesn't seem to be enough then you may be better off creating the brightness you need by using one of the enhancers currently on the market, such as an Aphex Exciter or an SPL Vitalizer. Those fortunate enough to have access to a parametric equaliser may prefer to use a broad-band boost. Bite can be added to the sound by boosting between 5kHz and 7kHz, while harshness in the upper mid range may be minimised by cutting between 1kHz and 3kHz. As usual, the exact frequency on which you will need to work is identified by setting the relevant equaliser section to full boost and then sweeping through the frequency range. When you hit it, the offending frequency should stand out like a sore thumb!

Only apply as much cut as you need to do the job or the result is likely to be quite unnatural. The only area in which excessive EQ might be justified is in taking out all of the bottom end to give a bright rhythm sound with as little body as possible so as not to compete with other instruments in the mix. By the same token, switching in your sub-bass hi-pass filter won't significantly alter the sound but it will keep unwanted low-frequency information to a bare minimum. Interestingly, the digital EQ implemented in some computer audio workstations and digital mixers may work better with acoustic guitars than traditional analogue equalisers, as they seem better able to adjust the tonal balance of a sound without the result sounding as though it's been equalised. Curiously, though, some digital EQs seem to require a lot more cut or boost to achieve the necessary subjective result than analogue equalisers do. Of course, every equaliser has its own character, so some experimentation will be necessary.

10 ELECTRIC GUITARS

The electric guitar sound is unusual in that it depends not only on the instrument itself but also on the amplification and the loudspeaker systems used. Add to this the effect of the room acoustics, the position of the speaker cabinet within that room and, most importantly, the way in which the instrument is played, and you can see why there are as many unique guitar sounds as there are guitarists.

Amplification

Guitar speaker systems tend to comprise ten- or twelve-inch drivers, either singly or in multiples, mounted in cabinets which may be fully sealed or open backed. Tweeters or mid-range speakers are, as a general rule, not employed. The distinctive overdrive sound is caused by harmonic distortion added in the amplifier, but fed through a full-range speaker system, such as a hi-fi, the resulting sound would be raspy and most unmusical. Guitar speakers have a poor high-frequency response and tend to become inefficient above 2 or 3kHz, which has the effect of filtering out the undesirable harmonics and resulting in a sound which still has plenty of edge or bite but doesn't sound buzzy or thin.

Many of the best-loved guitar speakers are designed so that the cones themselves add distortion when outputting at high power levels. If you hear someone talking about loudspeaker break-up modes they are referring to the way in which the sound is distorted at high power levels and not about driving a speaker to destruction!

Open-backed cabinets tend to have a fatter sound than closed-backed models, but the approach to miking them is similar. There is also a vast difference in sound between one make of amplifier and another, and most players still uphold the belief that valve amplifiers sound better than solid-state models.

Types Of Microphone

Because of the limited frequency response of guitar speaker systems, dynamic microphones are very often used to mic them up, and a general-purpose cardioid or omni will usually give quite adequate results. In situations where spill is a problem a cardioid mic would be the obvious choice, but an omni used at half the distance will result in roughly the same amount of spill, so the practical differences between the two mic types are not all that great.

Because guitar amplification systems aren't short on volume, you don't need to worry about mic sensitivity, and at the other end of the scale most good dynamics can handle the extreme SPLs produced by guitar amps.

Many American engineers still prefer to use a capacitor microphone on the electric guitar, and this helps to produce the American rock sound, which is not as fat as the British sound and has a more incisive edge. The bottom line is that, if you get the sound you want, you've used the right mic.

On some occasions a mic with a presence peak will help a sound cut through a mix more effectively, whereas an already aggressive sound might benefit if a fairly flat mic is used. At the bass end, the lowest string on a guitar sounds at a little under 100Hz, so an extended bass response is not really necessary. However, some amps produce a definite low-frequency clunk when the strings are picked hard, and this may produce a significant amount of low frequency, which a mic with a good low-end response should interpret rather better. This cabinet clunk is mainly featured in heavy rock guitar styles.

Environment

Guitar cabinets can either be close miked, miked from a distance or a combination of the two methods using two or more mics. In some ways you have to treat a guitar cabinet like an instrument in its own right because different sounds come from different parts of it. The majority of the sound may come from the speakers, but even with a sealed cabinet a significant amount of sound comes from the back and sides of the box, and with an open-backed cabinet as much sound escapes from the back as from the front.

The traditional way to get the sound of a cranked-up stack played live is to set up the full stack and then mike it from ten feet away or more in a large room. This way you will capture the direct sound from the speakers, including any phase-cancellation effects caused by multiple drivers, and you will also get the sound reflected from the floor, which creates further comb filtering. In other words, the mic hears the performance much as an audience would. However, few studios have the space to work in this way, especially when several members of the band are playing together, and some degree of acoustical separation is needed. Furthermore, although this gives a warm and loud quality of sound, it doesn't sound as bright or intimate as a close-miked amp and so may not cut through in a complex mix.

**Figure 10.1: Close-miked
guitar amplifier**

Mic pointed directly at
the speaker from a
position very close to the
grille cloth

A more basic approach is to position a single mic some
12 inches or less in front of one of the speakers in the
cabinet. Many engineers actually prefer to work with
the mic right up against the speaker cloth. If the mic is
pointed directly along the axis of the speaker then the
sound will be relatively bright, but it can be mellowed
by moving the mic towards the edge of the speaker.
Figure 10.1 shows a typical close-miking situation.

Open-Backed Speakers
With open-backed cabinets, a fatter sound may be
obtained by miking the rear of the cabinet, and some

engineers even like to mic the side of the cabinet and then add this sound to the direct miked sound. There is no reason not to mic both the front and rear of the cabinet simultaneously, but the phase of the rear mic should be inverted so that its output is in phase with that of the front mic.

The reason for this is pretty clear if you think about it. While the front of the speaker cone is travelling towards the front microphone, the rear of the cone must be travelling away from the rear mic at the same time, which will create an opposite polarity of signal. Without

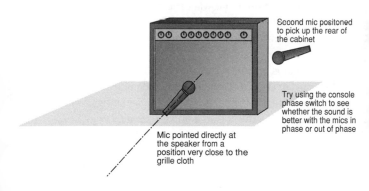

Second mic positoned to pick up the rear of the cabinet

Try using the console phase switch to see whether the sound is better with the mics in phase or out of phase

Mic pointed directly at the speaker from a position very close to the grille cloth

Figure 10.2: Miking the front and rear of the cabinet

the use of phase inversion, these two signals would cancel to some extent, giving a quite different and less weighty sound. Having said that, if you prefer the sound without using phase reversal, then feel free to use the sound you like! Figure 10.2 illustrates the dual-microphone approach.

Ambience

A popular alternative to the close-miked approach is to use an additional ambience mic, usually a capacitor model, several feet from the cabinet and then to add this to the close-miked sound. A direct comparison of the two sounds in isolation will reveal that the close-miked sound is more clinical, brighter and indeed sounds closer. The ambience mic will have a softer tone and will include a degree of room reverb, giving the sound more spread and making the tone smoother.

If the guitar is played in a room other than that in which the amplifier is located, then a capacitor mic can be used to capture the direct sound from the guitar strings, which can be added to the miked amplifier sound. On its own, the miked strings will seem thin and tinny, but when mixed in with the main sound they will add definition to the notes in the same manner as an exciter. This is definitely a technique with which it is worth experimenting, especially for country or clean Strat-type guitar sounds.

Equalisation

When it comes to the mix, some additional EQ may be needed to achieve the desired tone because the sound you hear in the studio is rarely exactly the same as the sound you get on tape. Much of the compensation can be carried out at the recording stage by adding EQ on the desk or the amplifier itself so that the sound heard over the monitors is more or less correct. Even so, once the other instruments are added the tone may need modifying to make it stand out against the rest of the mix.

If the sound is muddy, some cut applied at between 100Hz and 250Hz can help reduce the problem, and the usual method of determining the frequency at which to work still applies: crank up the low-mid boost and then sweep through the frequency range until the offending frequencies are most prominent. At that point you can change from boost to cut and set the amount of cut by ear.

Effects

It is often considered to be inadvisable to record effects at the same time as you record the original sounds because you lose the option to change the effects at the mixing stage, unless the effects can be recorded on separate tape tracks. However, with the guitar the effects are modified by the amplification system itself, and so it may be impossible to duplicate these later. Furthermore, the player responds to the sound he or she is producing, and the effects are often an integral part of the performance, particularly overdrive or delay. There are no hard and fast rules on which effects have to be recorded to tape and which you can add later, but it is important not to compromise the performance for the sake of some slight technical advantage. Discuss the options available and their implications with the player, and then try to find an agreed method of working with which you can both live.

Direct Injecting

It is possible to record the guitar without miking it up at all, and though direct injecting used to be the second-best option in recording situations some of today's digital recording guitar pre-amps based on physical modelling can sound superb. An advantage of direct injecting is that the guitarist can play in the control room and physically hear the actual sound which is going onto tape over the monitors as he or she plays. For clean guitar sounds, you don't even need a special recording pre-amp, but unfortunately, unless the electric guitar has active electronic components, it can't be plugged directly into the mixing console.

A simple DI box (a model with a high-impedance instrument input) will solve the impedance-matching problem, but the tone still won't quite be what you are used to as guitar amplifiers don't have a flat frequency response. Another factor to watch out for is the way in which the speaker changes the tonal character of the sound. A basic rhythm sound can be achieved by using a DI box and then applying some corrective or creative EQ, but a dedicated guitar processor will give better results.

Speaker Emulators

Another approach to direct-injected guitar is to use your regular amplifier and replace the speaker with a speaker emulator or simulator. These often take the form of a dummy load, allowing the amplifier to perform just as it would with a speaker, and a filtering network to emulate the coloration produced by a typical guitar speaker. The output signal is reduced to a level which the mixing console can accept, usually in the form of a balanced mic-level signal. Cheaper models are available without the dummy load, but that means that your guitar amp speakers must be left connected or you will have to use a separate dummy load. Most transistor amplifiers will work happily with the speakers unplugged, but valve amplifiers could sustain damage if used in this way.

Speaker simulators give a reasonably convincing rock sound in general, but they still don't quite capture the effect of loudspeaker break-up on the overdriven sound. The addition of a little reverb or ambience improves matters enormously, and then you also have the advantage of retaining the basic character of the amplifier being used.

Remiking

One method of working that has proven to be very useful is to treat DI'd sounds by feeding the guitar track through a guitar stack or combo during the mix, miking this up and then feeding the mic output back into the mixer. If the guitar is recorded unprocessed, this gives the engineer the option of producing virtually any type of sound subject only to the availability of a suitable guitar amp. It has been known for these to be set up in bathrooms or concrete stairwells to achieve a very convincing live sound. Of course there's no reason not to do this with a recording pre-amp, recording the guitar clean via a DI box and then patching in the pre-amp when you come to mix. In fact, these days the pre-amp doesn't even need to exist – it may come as a software plug-in for your computer audio system.

11 DRUMS

Unlike almost every other instrument, it's impossible to just take a drum kit that has been tuned to be used in a live environment and then expect it to sound equally good in a studio environment. In a live situation, some of the unwanted characteristics of the drum sound – the rattles, buzzes, metallic rings and that sort of thing – will go unnoticed or will be camouflaged by other instruments. In the studio, however, this noise shows up more clearly, and so some time must be spent in tuning and damping the kit before any microphones are set up and anything is recorded. It's also important not to forget to oil squeaking bass drum pedals and stools.

The drums should be set up in a slightly live environment so that they produce an exciting sound in the room. Some forms of music demand a very live environment, which usually means a large room with a lot of hard surfaces, but the smaller studio can also create some of this ambience by using artificial reverb at the mixing stage.

Single Heads

The easiest kit to record is the one that uses single-headed toms and a bass drum which has had a large hole cut from the front head. Some players simply remove the front head, but this can put uneven stress on the drum shell. The hole should be as large as is practical in order to prevent the remaining material from ringing, which could be picked up by the mics.

A wooden bass drum beater gives a better-defined sound than cork or felt beaters, and a patch of moleskin or hard plastic taped to the head at the point at which the beater connects will add more of a click to the sound. There are specialist drum products available for which can be used in this way.

Snares

Snare drums come in many sizes and are usually provided with metal or wooden shells. Wooden shells give a less ringing sound than metal ones, and a deeper shell generally equates to a deeper tone. Make sure that the snares themselves are in good order and properly adjusted to minimise rattling and buzzing.

All drummers have their own preferences for tuning, but generally speaking the snare head should be just slightly looser than the batter head. The heads should be tensioned evenly so that tapping the head around the edges produces a note of the same pitch all the way around the head. If the batter head is tensioned too tightly then the tone will become tubby and unpleasant. Conversely, if it is tuned too slackly, the sound will lack body. There is a range between these two extremes in which a variety of usable tones can be coaxed from any drum of reasonable quality.

Damping

A kit with no damping will usually sound very boomy and the individual drums will sustain for too long. Different heads will also affect the damping in different ways: a thin, hard head will give a bright tone with a long sustain, whereas thicker heads, or those utilising a double layer of material, will give a slightly fatter sound with a shorter decay, and some of the very heavy oil-filled heads sound just like hitting suitcases, even before damping is applied.

Too much damping can leave the kit sounding lifeless, and it's important to remember that the ringing and excessive sustain will sound less pronounced in the context of a mix, when other instruments are also playing.

Toms may be either single or double headed, and after being tuned and suitably damped they may be miked in much the same way as the snare drum. Toms may be damped by taping tissue or cloth to the heads near the edges, and it should be ensured that these damping pads aren't likely to be hit by the drummer. Similarly, try to keep the damping pad away from that area of head at which the mic is likely to be pointing.

The way in which toms are tuned affects their tone and decay characteristics, and a perfectly evenly tuned head may give the best sound. One popular trick is to tune the head evenly and then slacken off just one of the tuning lugs slightly. This gives a slight drop in pitch once the drum is struck, and is very popular in rock music.

Many drums are fitted with internal dampers, but unless these are quite sophisticated they are unlikely to provide good results. Gaffa tape, or pads of tissue taped to the head, produces a more even tone.

The snare drum may be damped in a similar way, but no damping should be applied to the snare head as this will dull the tone of the snares. As with the toms, don't overdo the tuning – it's best to err on the side of caution and add too little rather than too much.

Damping the bass drum is best achieved by placing a folded woollen blanket inside the drum so that it rests on the bottom of the shell and touches the lower part of the rear head. Further damping is unlikely to be necessary, though noise gates are often used to sharpen up the decay of the sound.

Miking Options

The simplest and perhaps most honest way to mic a drum kit is to position a stereo pair of microphones at around five feet from the ground and between five and 15 feet in front of the kit. Figure 11.1 over the page shows how this kind of arrangement may be set up in practice. These mics will capture the live sound and stereo imaging of the kit as played in that particular room very accurately, but they will also pick up spill from any other instruments that happen to be playing at the same time. More importantly, the accurate live sound of the kit is seldom the objective in pop music production.

The stereo pair approach is probably best suited to jazz work, but even here the snare and bass drums often benefit from a little extra help. To achieve this, close mics are used on the bass and snare drums and the stereo pair positioned several feet above the kit on extended boom stands. These overheads may be set up as a coincident pair, but it is more common to use spaced omnis. Capacitor mics are normally

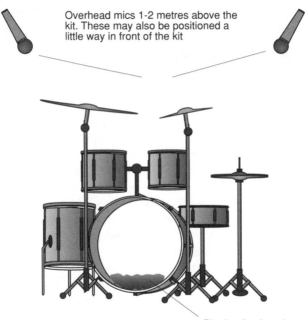

Overhead mics 1-2 metres above the kit. These may also be positioned a little way in front of the kit

Blanket for damping

Figure 11.1: Miking a drum kit using a stereo pair

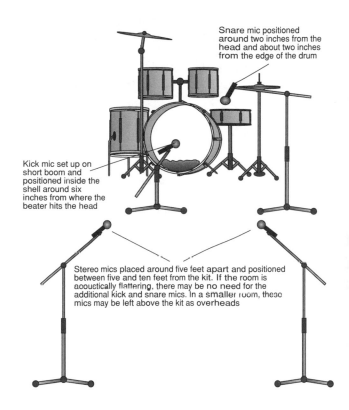

Snare mic positioned around two inches from the head and about two inches from the edge of the drum

Kick mic set up on short boom and positioned inside the shell around six inches from where the beater hits the head

Stereo mics placed around five feet apart and positioned between five and ten feet from the kit. If the room is acoustically flattering, there may be no need for the additional kick and snare mics. In a smaller room, these mics may be left above the kit as overheads

Figure 11.2: Overhead pair plus close bass and snare mics

used for this task as their high-end response is needed to faithfully reproduce the cymbals and the attack of the drums, although good results can be achieved by using budget PZM mics which use electret capsules. This setup is shown on the previous page in Figure 11.2.

Snare Mics

The snare mic may be either a dynamic or a capacitor type, depending on your preference and budget. Either variety will yield a useful sound, but the character from each will be different. As you might expect, the capacitor will give a crisper sound. Either directional or omni pattern mics may be used – in either case, the close proximity of other drums means that some spill is inevitable, and even though omnis will pick up a little more spill it will be tonally correct. Off-axis spill into a cardioid mic is likely to sound less bright than the on-axis sound.

The microphone should ideally be placed a couple of inches from the drum head and positioned to one side, where the drummer is unlikely to hit it. The mic may be tilted so that it points at the centre of the head, and it should be positioned so that it isn't pointing directly towards an adjacent drum, which would worsen the spill situation.

Bass Drum Mics

Bass drum mics are usually mounted on boom stands so that the mic can be positioned inside the drum shell. The exact position of the mic within the shell will influence the tonal character of the sound, and a good starting point is to have the mic pointing directly at the point on the head which the beater hits at a distance of between two and eight inches. Moving the mic to either side or angling it slightly will emphasise the overtones produced by the drum shell, so there is a lot of leeway with which to change the basic sound without having to resort to equalisation.

Large-diaphragm dynamic microphones tend to be favoured for bass drum work, and either figure-of-eight or cardioid models are the usual choice. Extremely high levels of sound pressure are generated inside bass drums, and so a mic capable of working at these levels must be used. Because of the low frequencies involved, a microphone with a good bass response is essential if you want the sound to have any body to it. Many specialist bass drum microphones, such as the AKG D12, have a low-frequency boost at around 80Hz to emphasise the thump of the drum, and their later D112 also includes a degree of high-end boost, which helps to accentuate the click of the beater striking the head.

Multi-Miking

When recording contemporary music, drums are nearly always miked individually, enabling the engineer to finely control the balance of the drums within the kit. Furthermore, close miking gives a more immediate sound, which is popular when a larger-than-life sound is required; and with pop or rock music, we nearly always want a drum sound more powerful than that produced by an acoustic kit played without amplification.

The approach to miking the bass, snare and overheads is the same as before, except that we must now add further mics to cover the toms. An additional mic may also be needed for the hi-hat, especially if the snare mic is going to be gated in order to improve separation.

Toms

The tom mics are positioned in a similar way to the snare mic, and may either be dynamic or capacitor types – cardioid or omni. A great many British engineers use dynamic cardioids, such as Shure SM57s or Sennheiser 421s, but there is a swing towards the use of capacitor mics because of their improved transient handling capability, which leads to a better-defined sound. The argument for using omnis rather than cardioids, so that any spill will at least sound tonally correct, is also valid.

The overheads and hi-hat mics should be capacitor mics in order to preserve the transient detail of the cymbals; but you may find a separate hi-hat mic unnecessary, depending on the setup of the kit, because the snare mic and overheads will pick up all the hi-hat level you will need. Take care in positioning separate hi-hat mics so that they don't get hit by a blast of air every time the hi-hats are closed, as this will spoil the sound. Positioning them a few inches from the edge of the cymbals and angling them from above or beneath should be adequate, but always experiment with different positions to find a better result. Figure 11.3 shows a fully miked kit, including hi-hat mic and overheads.

Percussion

Other forms of percussion, such as congas, may be miked from overhead in either mono or stereo, and unless separation is a problem, the microphone distance may be increased to anything between one and three feet from the drum head, depending on how much of the room ambience you wish to capture. If in doubt, keep in mind that positioning the mics close to the performer's head will capture more or less the same sound that they hear while playing. Additional damping is seldom required on Latin and ethnic percussion, but you should take care to find an acoustically sympathetic environment in which to work.

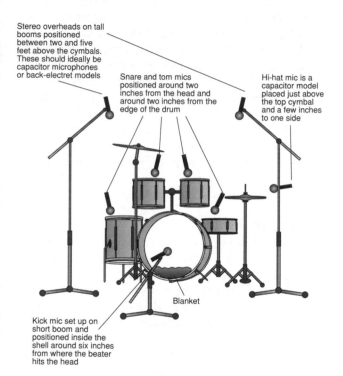

Stereo overheads on tall booms positioned between two and five feet above the cymbals. These should ideally be capacitor microphones or back-electret models

Snare and tom mics positioned around two inches from the head and around two inches from the edge of the drum

Hi-hat mic is a capacitor model placed just above the top cymbal and a few inches to one side

Blanket

Kick mic set up on short boom and positioned inside the shell around six inches from where the beater hits the head

Figure 11.3: A fully miked kit, including hi-hat mic and overheads

12 THE PIANO

With so many excellent and inexpensive MIDI piano modules currently on the market, you might wonder why anyone bothers to go to the trouble of recording the real thing – or at least you might if you hadn't already listened to the difference between a good piano and its sample-based emulation. The fact is that a piano is a living, breathing instrument full of resonances and vibrations that are far too complex to emulate with absolute accuracy. For example, whenever you play a note on a real piano, the other strings also vibrate in sympathy, but in different ways, depending on which note you've played and how hard you've played it. If you play two notes at a time the pattern of sympathetic resonances gets more complex, but because of the way in which most electronic pianos are sampled the best you can hope for is that each note will be accompanied by the sympathetic resonances that occur when that note is played in isolation. And unless each separate note is sampled (as opposed to one note being sampled and then transposed to cover several keys), the resonances will also be transposed.

There's also the way that the timbre of a real piano changes with dynamics. Velocity crossfading between a few discrete samples or using a velocity-controlled filter is never going to capture the subtle nuances of a real piano played by a virtuoso performer and, while a MIDI piano might be fine as part of a mix, few piano players feel entirely comfortable using them for solo performances or prominent parts. In such cases, there's no alternative but to get out the mics and record the real thing.

The Sound Of The Piano

When it comes to recording, the acoustic piano isn't without its problems – like most acoustic instruments, the sound doesn't just emanate from one convenient point but from a combination of the strings, the soundboard and the casework of the instrument. It's only when all of these vibrating parts make their contribution that a true sound can be captured. To frustrate the recording engineer even further, there are various mechanical noises that must be minimised, such as those produced by the pedals and dampers, and while a change in playing style is sometimes all that's needed it's not uncommon to have to wrap the pedals in cloth to stop them from thumping. And of course, unlike MIDI, you also have acoustic spill and room acoustics to worry about.

The most important piano recordings are made using a grand piano, unless the musical style specifically requires the sound of an upright, and it's generally thought that the larger the piano the better the sound, especially at the bass end. Before investigating mic positions, it's necessary to think about the sound of the instrument and the types of mic that will be needed to do that sound justice. The piano spans the entire musical spectrum, from deep bass to almost ultrasonic harmonics, so a microphone with a wide-frequency response is a must. A good capacitor or back-electret microphone is the preferred choice, though you can exercise a degree of artistic choice in deciding to go for a ruthlessly honest small-capsule model or a more flattering but less accurate large-diaphragm mic. Along with their excellent frequency response capacitor mics are also very sensitive, which means that they'll be able to capture the dynamic range of the instrument without introducing unwanted noise, even when you're recording from several feet away. For demo work, good dynamic mics will produce acceptable results, but capacitor models are really the only choice for release-quality recordings.

Pianos are most often recorded in stereo, and any of the standard stereo-mic techniques may be applied. You can use spaced omnis, coincident cardioids, MS pairs or PZM mics, though if you're worried about mono compatibility you may feel safer sticking to a coincident mic setup rather than a spaced arrangement, as spaced mics, by their very geometry, introduce phase effects which may cause the sound to change for the worse when you hit the console's mono button. On the other hand, spaced microphones allow more control in balancing the upper and lower registers of the instrument, especially if you opt for a close-miking approach.

Some engineers habitually use three or more mics, but if you use this approach you run the risk of encountering more serious phase problems, so unless you've got plenty of time to experiment it may be safer to stick with a stereo pair. The only exception to this rule is when using spaced omnis some distance from the piano, as this often leads to a lack of definition in the centre of the soundstage. Positioning an additional centre mic can help to hold the sound together.

Mic Positions

Before deciding on a mic's position, you should first decide on the sound you want. Pop recordings may require a bright, up-front sound where being accurate is less important than getting the right type of sound, while for classical and jazz work you will probably want to capture the resonance of the instrument as accurately as possible, along with a little ambience from the environment.

Assuming that the room is sympathetic to the piano and that spill from nearby instruments isn't a problem, a simple stereo pair positioned between six and ten feet from the right hand (opening) side of the piano may be all that is needed. If the sound is becoming clouded by room ambience you can move in closer, whereas if the room is making a positive contribution to the sound you can afford to move the mics further out. Choosing cardioid mics means that you can work further from the piano than with omnis for the same amount of spill or room ambience, so if the room really isn't helping the sound then using cardioids or even hypercardioids might help.

Keep the microphones at such a height that they're aimed about halfway up the inside of the open lid. If the acoustics of the room require that the microphones be brought in very close, it may be advantageous to add a little artificial reverb afterwards. This should be used sparingly, however, because the sympathetic resonances of the piano's own strings and soundboard provide a type of reverb. Figure 12.1 shows a suggested position for both coincident and spaced mics.

Spaced or coincident microphone pair placed between six and ten feet from the piano. Spaced mics may be around six feet apart, but experiment to find the best result

Piano with open side facing microphones

Figure 12.1: Stereo-miking a grand piano

For pop work, close miking is often employed to create a more cutting 'in your face', sound and one popular technique is to position the microphones inside the open piano lid between six and ten inches from the strings, close to the spot where the hammers strike the strings. Though cardioids may be used, omnis might be a better choice as they have a more accurate off-axis response, which tends to produce a more even tone across the strings. One covers the higher octaves and the other the bass end, and although this doesn't produce an accurate stereo recording it does sound good in stereo. Having said that, it may not be wise to pan the two mics to their extremes, as you could end up with a piano that sounds about 20 feet wide! Trust your ears on this one and use just enough panning to create a convincing sound.

Close miking has the additional benefit of reducing the amount of spill from other instruments, and if even greater separation is necessary then blankets may be draped over the piano lid to cover the opening. With any method involving spaced mics, whether close or distant, it's a good idea to perform a mono compatibility check by running through scales, using the whole keyboard. Any unduly loud or quiet notes or groups of notes, which may be attributable to either phase addition or cancellation, should be quite evident; and if the mono compatibility has suffered too much, try changing the spacing between the mics slightly and ensure that there is at least three times (and ideally five times) the distance between the two mics as there is between the mics and the strings. Figure 12.2 shows a typical close-miking arrangement.

Having suggested a few microphone positions based on experience and common practice, don't be afraid to experiment – nearly all initial microphone setups

basic Microphones

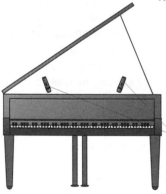

Mics may be either omni or cardioid, though omnis are likely to produce a more even sound. If coincident cardioid mics are used, these should be placed centrally and then moved slightly if the balance across the keyboard is uneven

Figure 12.2: Close-miking a grand piano

Position the mics between six and twelve inches from the strings. If the mics are too close, they will over-emphasise the strings nearest to them

148

can be improved upon or fine tuned simply by trying new positions. For example, if you need a more mellow tone or a fuller sound, consider putting a mic under the piano to capture more of the sound coming from the soundboard. Similarly, if the room is too live or coloured, experiment with improvising screens around the piano by using bedding or sleeping bags.

Boundary Mics

Some engineers like to use PZMs or boundary microphones for the recording piano, and the underside of the piano lid makes a perfect baffle. Simply tape the two mics to the underside of the lid with one mic favouring the high strings and the other the low strings. Be careful to use adhesive tape that can be removed without damaging the finish, however, especially if you're dealing with a concert grand! Though the popular Tandy/Radio Shack PZM mics don't turn in anything like the performance of professional studio PZMs or boundary microphones, they can yield unexpectedly good results when used in this way.

Upright Pianos

Though this description is likely to offend the purists, when it comes to miking it up the upright piano can essentially be thought of as approximating a grand piano turned up on its endpoint. The soundboard is now situated at the back of the instrument rather than underneath it, and because there's no single large lid covering all of the strings the only way in which it can be properly close-miked is by removing some or all of the front panelling to expose the strings.

The upright piano has a noticeably different tone to the larger grand, especially in the bass register, where the sound it produces can be much less rich, but with a little perseverance it is still possible to get a good miked-up sound.

A stereo pair of either coincident or spaced microphones can be positioned on boom stands above the instrument with the top of the piano left open. Check for an even tonal balance right across the range of the instrument and move the mics if there are any obvious dead or hot spots. Spaced mics are easier to use as they can be moved independently to balance the high and low ends of the instrument, but the down side is, as ever, the potential for audible phase problems. Figure 12.3 shows a typical miking arrangement.

Standing the piano close to a solid wall can help to beef up the bass end by getting the boundary effect on your side, and if the room suits the piano sound you're after then you can move the mics back to a few

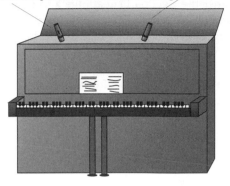

Mics should be facing into the open top of the piano. However, if it is possible to remove the front covers as well, this may help produce a bigger sound. Adjust the mic positions to get an even level across the strings

Upright piano

Figure 12.3: Stereo-miking an upright piano

feet behind the player. A room with a tiled, wooden or stone floor will help maintain a bright, lively tone, whereas carpet will tend to rob the sound of some of its sparkle. If you're struggling to get a mic position that works, it's always worth remembering that an instrument nearly always sounds OK to the person playing it, so try setting up the mics to 'look' over his or her shoulders.

basic Microphones

When close-miking the upright piano so that the mics are pointing directly at the strings, you can choose to mic either the section beneath the keyboard or the section directly above it. The most important things to do are to get an even level across the full range of the keyboard and to check for mono compatibility.

Because upright pianos are often set up in rooms which are acoustically imperfect, and because the large majority of them are instruments of indifferent quality, a little equalisation may prove useful in shaping the sound to your needs. It's best to try and keep the level of EQ to a minimum, however, and be careful to use gentle slopes or cuts rather than harsh boosts. To add sparkle, try using a little boost at around 6kHz or use an exciter/enhancer sparingly. An indifferent upright piano played in a carpeted room may benefit from both high-end EQ and a little artificial ambience or reverb, of which plate settings are the most flattering.

There's no real black art to recording the piano – you just need to apply a little logic and be prepared to experiment. If you're operating on a tight budget, a couple of cheap PZMs (sadly now discontinued) from your local Tandy/Radio Shack will work far better than could otherwise be reasonably expected, considering their price, and with the current preponderance of good-quality, inexpensive capacitor microphones, even doing the job seriously isn't prohibitively expensive. Don't be afraid to give it a try – provided that the microphones are half-decent and the piano is in tune, you're almost certain to end up with a usable recording.

The Piano As Reverb

As a novel alternative to reverb, try using an effects send to drive a small instrument amplifier or hi-fi amplifier and speaker placed underneath a grand piano or behind an upright piano. Close-mic the piano strings, jam down the sustain pedal using a convenient brick, and use the sympathetic resonance of the strings as a substitute for reverb. This method works particularly well when recording plucked or percussive sounds. Similarly, the recording of an electronic piano can be played back into a real piano through a loudspeaker, which will add a little sympathetic vibration.

13 STRINGED INSTRUMENTS

When recording strings, the choice of mic pattern is determined largely by the acoustics of the studio or venue and the need to minimise spill, and for this reason cardioids are often used. However, as in most other applications, a good omni will give the most natural-sounding results – as long as the room acoustics and spill considerations allow it.

Single Instruments

A solo violin, viola or cello may be miked using a single mic positioned three to five feet from the soundboard and pointing directly towards the instrument. Because of the positions of violins and violas, the mic will need to be set up on a tall stand or suspended from the ceiling. Don't place the mic too close to the ceiling, though, as reflections may be picked up.

Cellos and double basses may be miked using a short floor stand or a boom folded back on itself, and for a double bass a microphone with a frequency response extending down to 40Hz or below is necessary if the

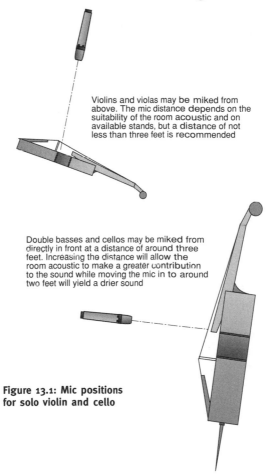

Violins and violas may be miked from above. The mic distance depends on the suitability of the room acoustic and on available stands, but a distance of not less than three feet is recommended

Double basses and cellos may be miked from directly in front at a distance of around three feet. Increasing the distance will allow the room acoustic to make a greater contribution to the sound while moving the mic in to around two feet will yield a drier sound

Figure 13.1: Mic positions for solo violin and cello

low notes are to be recorded faithfully. Figure 13.1 shows how the mics might be arranged in practice.

Rock 'n' roll double bass often lacks power when miked conventionally, so attaching a transducer pickups attached to the bridge is quite common. The sound of the pickup will generally be added to the miked sound, but check both positions of the mic phase switch to see which produces the most bass. Alternatively, a good dynamic mic wrapped in cloth and placed inside the instrument via the sound-hole will often do the trick. In this case, a degree of EQ may be needed to obtain a satisfactory tone, and so an external graphic or parametric equaliser should be patched in if available.

Multiple Instruments

Individual miking isn't necessary for larger string ensembles, and even for smaller ensembles, such as quartets, one of the stereo-mic techniques discussed elsewhere in this book will yield the best results in a suitable room. For example, a stereo pair of mics placed a few feet above or in front of the ensemble will usually work well. If the acoustic is too dry, some artificial reverberation may be added at the mixing stage.

Stereo-pair recording is only suitable for recording large ensembles in acoustically favourable venues because of the high proportion of ambient sounds. In this case, the engineer must monitor the output from the mics and adjust their distance from the musicians until the balance of direct and ambient sound is considered correct. In some cases it may be necessary to move whole groups of musicians backwards or forwards if they are too loud or too quiet in the overall mix.

Harps

Harps produce a very low acoustic output, so the distance at which you can work is limited mainly by mic noise and spill considerations. A cardioid capacitor mic perpendicular to and about two feet from the sound-board should give a fairly accurate result, and if the mic's close proximity makes the sound bass-heavy then a little low EQ cut may be necessary to compensate. The mic should 'look' through the strings towards the soundboard and be positioned at the opposite side of the instrument to the player.

Other plucked string instruments may be miked in a similar way to the acoustic guitar.

APPENDIX
Common Cable Connections

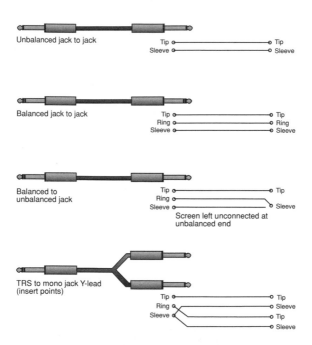

Unbalanced jack to jack

Tip o————————o Tip
Sleeve o————————o Sleeve

Balanced jack to jack

Tip o————————o Tip
Ring o————————o Ring
Sleeve o————————o Sleeve

Balanced to
unbalanced jack

Tip o————————o Tip
Ring o
Sleeve o————————o Sleeve

Screen left unconnected at
unbalanced end

TRS to mono jack Y-lead
(insert points)

Tip o————————o Tip
Ring o————————o Sleeve
Sleeve o————————o Tip
o Sleeve

Appendix: Common Cable Connections

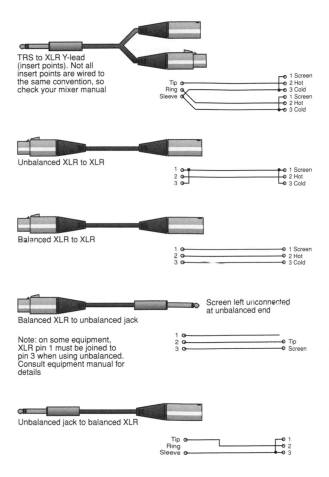

TRS to XLR Y-lead
(insert points). Not all
insert points are wired to
the same convention, so
check your mixer manual

Tip
Ring
Sleeve

1 Screen
2 Hot
3 Cold
1 Screen
2 Hot
3 Cold

Unbalanced XLR to XLR

1
2
3

1 Screen
2 Hot
3 Cold

Balanced XLR to XLR

1
2
3

1 Screen
2 Hot
3 Cold

Balanced XLR to unbalanced jack

Screen left unconnected
at unbalanced end

Note: on some equipment,
XLR pin 1 must be joined to
pin 3 when using unbalanced.
Consult equipment manual for
details

1
2
3

Tip
Screen

Unbalanced jack to balanced XLR

Tip
Ring
Sleeve

1
2
3

GLOSSARY

AC
Alternating Current.

Active
Circuit containing transistors, ICs, tubes and other devices that require power to operate and are capable of amplification.

A/D Converter
Circuit for converting analogue waveforms into a series of values represented by binary numbers. The more bits a converter has, the greater the resolution of the sampling process. Current effects units are generally 16 bits or more, with the better models being either 20- or 24-bit.

Ambience
The result of sound reflections in a confined space being added to the original sound. Ambience may also be created electronically by some digital reverb units. The main difference between ambience and

reverberation is that ambience doesn't have the characteristic long delay time of reverberation – the reflections mainly give the sound a sense of space.

Amp
Unit of electrical current, short for *ampère*.

Amplifier
Device that increases the level of an electrical signal.

Amplitude
Another word for level. Can refer to levels of sound or electrical signal.

Analogue
Describes circuitry that uses a continually changing voltage or current to represent a signal. The origin of the term is that the electrical signal can be thought of as being analogous to the original signal.

ASCII
American Standard Code for Information Interchange. A standard code for representing computer keyboard characters with binary data.

Attack
Time taken for a sound to achieve maximum amplitude.

Drums have a fast attack, whereas bowed strings have a slow attack. In compressors and gates, the attack time equates to how quickly the processor can change its gain.

Attenuate

To make lower in level.

Audio Frequency

Signals in the human audio range, nominally 20Hz–20kHz.

Balance

This word has several meanings in recording. It may refer to the relative levels of the left and right channels of a stereo recording, or it may be used to describe the relative levels of the various instruments and voices within a mix.

Bandpass Filter

Filter that removes or attenuates frequencies above and below the frequency at which it is set. Frequencies within the band are emphasised. Bandpass filters are often used in synthesisers as tone-shaping elements.

Bandwidth

Means of specifying the range of frequencies passed by an electronic circuit such as an amplifier, mixer or filter.

The frequency range is usually measured at the points where the level drops by 3dB relative to the maximum.

Boost/Cut Control
Single control which allows the range of frequencies passing through a filter to be either amplified or attenuated. The centre position is usually the 'flat' or 'no effect' position.

Bouncing
Process of mixing two or more recorded tracks together and re-recording these onto another track.

Buffer
Circuit designed to isolate the output of a source device from loading effects due to the input impedance of the destination device.

Buss
Common electrical signal path along which signals may travel. In a mixer, there are several busses carrying the stereo mix, the groups, the PFL signal, the aux sends and so on. Power supplies are also fed along busses.

Byte
Piece of digital data comprising eight bits.

Cardioid
Literally 'heart-shaped'. Describes the polar response of a unidirectional microphone.

Channel
Single strip of controls in a mixing console relating to either a single input or a pair of main/monitor inputs.

Channel
In the context of MIDI, Channel refers to one of 16 possible data channels over which MIDI data may be sent. The organisation of data by channels means that up to 16 different MIDI instruments or parts may be addressed using a single cable.

Channel
In the context of mixing consoles, a channel is a single strip of controls relating to one input.

Chip
Integrated circuit.

Chord
Two or more different musical notes played at the same time.

Chorus

Effect created by doubling a signal and adding delay and pitch modulation.

Click Track
Regular metronome pulse which helps musicians to keep time.

Clipping
Severe form of distortion which occurs when a signal attempts to exceed the maximum level which a piece of equipment can handle.

Compander
Encode/decode device that compresses a signal while encoding it, then expands it when decoding it.

Compressor
Device designed to reduce the dynamic range of audio signals by reducing the level of high signals or by increasing the level of low signals.

Computer
Device for the storing and processing of digital data.

Conductor
Material that provides a low resistance path for electrical current.

Console
Alternative term for mixer.

Contact Enhancer
Compound designed to increase the electrical conductivity of electrical contacts such as plugs, sockets and edge connectors.

Copy Protection
Method used by software manufacturers to prevent unauthorised copying.

Crash
Term relating to the malfunction of a computer program.

Cut-And-Paste Editing
Copying or moving sections of a recording to different locations.

Cutoff frequency
Frequency above or below which attenuation begins in a filter circuit.

Cycle
A complete vibration of a sound source or its electrical equivalent. One cycle per second is expressed as 1Hz (Hertz).

CV

Control Voltage. Used to control the pitch of an oscillator or filter frequency in an analogue synthesiser. Most analogue synthesisers follow a one volt per octave convention, though there are exceptions. To use a pre-MIDI analogue synthesiser under MIDI control, a MIDI-to-CV converter is required.

Daisy Chain

Term used to describe serial electrical connection between devices or modules.

Damping

In the context of reverberation, damping refers to the rate at which reverberant energy is absorbed by the various surfaces in an environment.

DAT

Digital Audio Tape. The most commonly used DAT machines are more correctly known as R-DATs because they utilise a rotating head similar to those found in video recorders. Digital recorders that use fixed or stationary heads (such as DCC) are known as S-DAT machines.

Data

Information stored and used by a computer.

Data Compression

System for reducing the amount of data stored by a digital system. Most audio data compression systems are known as lossy systems, as some of the original signal is discarded in accordance with psychoacoustic principles designed to ensure that only components which cannot be heard are lost.

dB

Decibel. Unit used to express the relative levels of two electrical voltages, powers or sounds.

dBm

Variation on dB referenced to 0dB = 1mW into 600 ohms.

dBv

Variation on dB referenced to 0dB = 0.775v.

dBV

Variation on dB referenced to 0dB = 1V.

dB Per Octave

A means of measuring the slope of a filter. The more decibels per octave the sharper the filter slope.

dbx

A commercial encode/decode tape noise-reduction system

that compresses the signal during recording and expands it by an identical amount on playback.

DC
Direct Current.

DCO
Digitally Controlled Oscillator.

DDL
Digital Delay Line.

Decay
Progressive reduction In amplitude of a sound or electrical signal over time. In an ADSR envelope shaper, the decay phase starts as soon as the attack phase has reached its maximum level. In the decay phase, the signal level drops until it reaches the sustain level set by the user. The signal then remains at this level until the key is released, at which point the release phase is entered.

De-esser
Device for reducing the effect of sibilance in vocal signals.

Deoxidising Compound
Substance formulated to remove oxides from electrical contacts.

Detent
Physical click stop in the centre of a control such as a pan or EQ cut/boost knob.

DI
Direct Inject, in which a signal is plugged directly into an audio chain without the aid of a microphone.

DI Box
Device for matching the signal-level impedance of a source to a tape machine or mixer input.

Digital
Describes an electronic system which represents data and signals in the form of codes comprising 1s and 0s.

Digital Delay
Digital processor for generating delay and echo effects.

Digital Reverb
Digital processor for simulating reverberation.

DIN Connector
Consumer multi-pin signal connection format, also used for MIDI cabling. Various pin configurations are available.

Direct Coupling

Means of connecting two electrical circuits so that both AC and DC signals may be passed between them.

Disc

Term used to describe vinyl record discs, CDs and MiniDiscs.

Dither

System of adding low-level noise to a digitised audio signal in such a way that extends the low-level resolution at the expense of a slight deterioration in noise performance.

Dolby

An encode/decode tape noise reduction system that amplifies low-level, high-frequency signals during recording, then reverses this process during playback. There are several different Dolby systems in use, including types B, C and S for domestic and semi-professional machines, and types A and SR for professional machines. Recordings made whilst using one of these systems must also be replayed via the same system.

DOS

Disk Operating System. Part of the operating system of PC and PC-compatible computers.

Driver

Piece of software that handles communications between the main program and a hardware peripheral, such as a soundcard, printer or scanner.

Drum Pad

Synthetic playing surface which produces electronic trigger signals in response to being hit with drumsticks.

Dry

Signal to which no effects have been added. Conversely, a sound which has been treated with an effect, such as reverberation, is referred to as wet.

DSP

Digital Signal Processor. A powerful microchip used to process digital signals.

Dubbing

Adding further material to an existing recording. Also known as overdubbing.

Ducking

System for controlling the level of one audio signal with another. For example, background music can be made to duck whenever there is a voice-over.

Dynamic Microphone

Type of microphone that works on the electric generator principle, whereby a diaphragm moves a coil of wire within a magnetic field.

Dynamic Range

Range in decibels between the highest signal that can be handled by a piece of equipment and the level at which the smallest signals disappear into the noise floor.

Dynamics

Method of describing the relative levels within a piece of music.

Early Reflections

First sound reflections from walls, floors and ceilings following a sound which is created in an acoustically reflective environment.

Effects Loop

Connection system that allows an external signal processor to be connected into the audio chain.

Effects Return

Additional mixer input designed to accommodate the output from an effects unit.

Effects Unit

Device for treating an audio signal in order to change it in some creative way. Effects often involve the use of delay circuits, and include such treatments as reverb and echo.

Encode/Decode

System that requires a signal to be processed prior to recording, which is then reversed during playback.

Enhancer

Device designed to brighten audio material using techniques such as dynamic equalisation, phase shifting and harmonic generation.

Envelope

The way in which the level of a sound or signal varies over time.

Envelope Generator

Circuit capable of generating a control signal which represents the envelope of the sound you want to recreate. This may then be used to control the level of an oscillator or other sound source, though envelopes may also be used to control filter or modulation settings. The most common example is the ADSR generator.

Equaliser

Device for selectively cutting or boosting selected parts of the audio spectrum.

Erase

To remove recorded material from an analogue tape, or to remove digital data from any form of storage medium.

Exciter

Enhancer that works by synthesising new high-frequency harmonics.

Expander

Device designed to decrease the level of low-level signals and increase the level of high-level signals, thus increasing the dynamic range of the signal.

Expander Module

Synthesiser with no keyboard, often rack mountable or in some other compact format.

Fader

Sliding control used in mixers and other processors.

Figure-Of-Eight

Describes the polar response of a microphone that is

equally sensitive at both front and rear, yet rejects sounds coming from the sides.

File
Meaningful list of data stored in digitally. A Standard MIDI File is a specific type of file designed to allow sequence information to be exchanged between different types of sequencer.

Filter
Electronic circuit designed to emphasise or attenuate a specific range of frequencies.

Flanging
Modulated delay effect using feedback to create a dramatic, sweeping sound.

Flutter Echo
Resonant echo that occurs when sound reflects back and forth between two parallel reflective surfaces.

Foldback
System for feeding one or more separate mixes to the performers for use while recording and overdubbing. Also known as a *cue mix*.

Format

Procedure required to ready a computer disk for use. Formatting organises the disk's surface into a series of electronic pigeonholes into which data can be stored. Different computers often use different systems of formatting.

Frequency
Indication of how many cycles of a repetitive waveform occur in one second. A waveform which has a repetition cycle of once per second has a frequency of 1Hz.

Frequency Response
Measurement of the frequency range that can be handled by a specific piece of electrical equipment or loudspeaker.

Fundamental
Any sound comprises a fundamental or basic frequency plus harmonics and partials at a higher frequency.

FX
Shorthand for effects.

Gain
Amount by which a circuit amplifies a signal.

Gate
Electrical signal that is generated when a key is

depressed on an electronic keyboard. Used to trigger envelope generators and other events that need to be synchronised to key action.

Gate
Electronic device designed to mute low-level signals, thus improving the noise performance during pauses in the wanted material.

General MIDI
Addition to the basic MIDI spec to assure a minimum level of compatibility when playing back GM-format song files. The specification covers type and program, number of sounds, minimum levels of polyphony and multitimbrality, response to controller information and so on.

Glitch
Describes an unwanted short-term corruption of a signal, or the unexplained short-term malfunction of a piece of equipment. For example, an inexplicable click on a DAT tape would be termed a glitch.

Graphic Equaliser
Equaliser on which several narrow segments of the audio spectrum are controlled by individual cut/boost faders. The name of the device derives from the fact

that the fader positions provide a graphic representation of the EQ curve.

Ground

Electrical earth, or zero volts. In mains wiring, the ground cable is physically connected to the ground via a long conductive metal spike.

Ground Loops

Also known as earth loops. Wiring problem in which currents circulate in the ground wiring of an audio system, known as the ground loop effect. When these currents are induced by the alternating mains supply, hum results.

Group

Collection of signals within a mixer that are mixed and then routed through a separate fader to provide overall control. In a multitrack mixer, several groups are provided to feed the various recorder track inputs.

Harmonic

High-frequency component of a complex waveform.

Harmonic Distortion

Addition of harmonics not present in the original signal.

Head
Part of a tape machine or disk drive that reads and/or writes data to and from the storage media.

Headroom
The safety margin in decibels between the highest peak signal being passed by a piece of equipment and the absolute maximum level the equipment can handle.

High-Pass Filter
Filter which attenuates frequencies below its cutoff frequency.

Hiss
Noise caused by random electrical fluctuations.

Hum
Signal contamination caused by the addition of low frequencies, usually related to the mains power frequency.

Hz
Shorthand for Hertz, the unit of frequency.

IC
Integrated Circuit.

Impedance

Can be visualised as the AC resistance of a circuit which contains both resistive and reactive components.

Inductor

Reactive component which presents an impedance with increases with frequency.

Insert Point

Connector that allows an external processor to be patched into a signal path so that the signal then flows through the external processor.

Insulator

Material that does not conduct electricity.

Interface

Device that acts as an intermediary to two or more other pieces of equipment. For example, a MIDI interface enables a computer to communicate with MIDI instruments and keyboards.

Intermittent

Usually describes a fault that only appears occasionally.

Intermodulation Distortion

Form of distortion that introduces frequencies not present

in the original signal. These are invariably based on the sum and difference products of the original frequencies.

I/O
The part of a system that handles inputs and outputs, usually in the digital domain.

Jack
Common audio connector. May be mono (TS) or stereo (TRS).

k
Abbreviation of 1,000 (kilo). Used as a prefix to other values to indicate magnitude.

Limiter
Device that controls the gain of a signal so as to prevent it from ever exceeding a preset level. A limiter is essentially a fast-acting compressor with an infinite compression ratio.

Linear
Describes a device from which the output is a direct multiple of the input.

Line Level
Mixers and signal processors tend to work at a standard

signal level known as line level. In practice there are several different standard line levels, but all are in the order of a few volts. A nominal signal level is around −10dBv for semi-pro equipment and +4dBv for professional equipment.

Load

Electrical circuit that draws power from another circuit or power supply. Also describes reading data into a computer.

Loop

Circuit where the output is connected back to the input.

Low-Pass Filter

Filter which attenuates frequencies above its cutoff frequency.

mA

Milliamp, or one thousandth of an amp.

MDM

Modular Digital Multitrack. A digital recorder that can be used in multiples to provide a greater number of synchronised tracks than a single machine.

Mic Level

Low-level signal generated by a microphone. This must be amplified many times to increase it to line level.

MIDI
Musical Instrument Digital Interface.

MIDI Bank Change
Type of controller message used to select alternate banks of MIDI programs where access to more than 128 programs is required.

MIDI Controller
Term used to describe the physical interface by means of which the musician plays the MIDI synthesiser or other sound generator. Examples of controllers are keyboards, drum pads, wind synths and so on.

MIDI Control Change
MIDI message also known as MIDI controllers or controller data. These messages convey positional information relating to performance controls such as wheels, pedals, switches and other devices. This information can be used to control functions such as vibrato depth, brightness, portamento, effects levels, and many other parameters.

(Standard) MIDI File
Standard file format for storing song data recorded on

a MIDI sequencer in such as way as to allow it to be read by other makes or models of MIDI sequencer.

MIDI In
The socket used to receive information from a master controller or from the MIDI Thru socket of a slave unit.

MIDI Merge
Device or sequencer function that enables two or more streams of MIDI data to be combined.

MIDI Mode
MIDI information can be interpreted by the receiving MIDI instrument in a number of ways, the most common being polyphonically on a single MIDI channel (Poly [Omni Off] mode). Omni mode enables a MIDI Instrument to play all incoming data regardless of channel.

MIDI Module
Sound-generating device with no integral keyboard.

MIDI Note Number
Every key on a MIDI keyboard has its own note number, ranging from 0 to 127, where 60 represents middle C. Some systems use C3 as middle C while others use C4.

MIDI Note Off

MIDI message sent when key is released.

MIDI Note On

Message sent when note is pressed.

MIDI Out

MIDI connector used to send data from a master device to the MIDI In of a connected slave device.

MIDI Port

MIDI connections of a MIDI-compatible device. A multiport, in the context of a MIDI interface, is a device with multiple MIDI output sockets, each capable of carrying data relating to a different set of 16 MIDI channels. Multiports are the only means of exceeding the limitations imposed by 16 MIDI channels.

MIDI Program Change

Type of MIDI message used to change sound patches on a remote module or the effects patch on a MIDI effects unit.

MIDI Sync

Description of the synchronisation systems available to MIDI users: MIDI Clock and MIDI Time Code.

MIDI Thru

Socket on a slave unit used to feed the MIDI In socket of the next unit in line.

MIDI Thru Box

Device which splits the MIDI Out signal of a master instrument or sequencer to avoid daisy chaining. Powered circuitry is used to 'buffer' the outputs so as to prevent problems when many pieces of equipment are driven from a single MIDI output.

Mixer

Device for combining two or more audio signals.

Monitor

Reference loudspeaker used for mixing.

Monitor

VDU for a computer.

Monitoring

Action of listening to a mix or a specific audio signal.

Monophonic

One note at a time.

MTC

MIDI Time Code. A MIDI sync implementation based on SMPTE time code.

Multitimbral Module

MIDI sound source capable of producing several different sounds at the same time and controlled on different MIDI channels.

Multitrack

Recording device capable of recording several 'parallel' parts or tracks which may then be mixed or re-recorded independently.

Near Field

Some people prefer the term 'close field' to describe a loudspeaker system designed to be used close to the listener. The advantage is that the listener hears more of the direct sound from the speakers and less of the reflected sound from the room.

Noise Reduction

System for reducing analogue tape noise or for reducing the level of hiss present in a recording.

Normalise

A socket is said to be normalised when it is wired such that the original signal path is maintained, unless a plug is inserted into the socket. The most common examples of normalised connectors are the insert points on a mixing console.

Octave
When a frequency or pitch is transposed up by one octave, its frequency is doubled.

Offline
Describes a process carried out while a recording is not playing. For example, some computer-based processes have to be carried out offline as the computer isn't fast enough to carry out the process in real time.

Ohm
Unit of electrical resistance.

Omni
Abbreviation of *omnidirectional*. Refers to a microphone that is equally sensitive in all directions, or to the MIDI mode in which data on all channels is recognised.

Oscillator
Circuit designed to generate a periodic electrical waveform.

Overdubbing

Adding another part to a multitrack recording or to replace one of the existing parts. (see *Dubbing*.)

Overload
To exceed the operating capacity of an electronic or electrical circuit.

Pad
Resistive circuit for reducing signal level.

Pan Pot
Control enabling the user of a mixer to move the signal to any point in the stereo soundstage by varying the relative levels fed to the left and right stereo outputs.

Parallel
Method of connecting two or more circuits together so that their inputs and outputs are all connected together.

Parameter
Variable value that affects some aspect of a device's performance.

Parametric EQ
Equaliser with separate controls for frequency, bandwidth and cut/boost.

Passive

Describes a circuit with no active elements.

Patch

Alternative term for program. Referring to a single programmed sound within a synthesiser that can be called up using program-change commands. MIDI effects units and samplers also have patches.

Patch Bay

System of panel-mounted connectors used to bring inputs and outputs to a central point from where they can be routed using plug-in patch cords.

Patch Cord

Short cable used with patch bays.

Peak

Maximum instantaneous level of a signal. The highest signal level in any section of programme material.

PFL

Pre-Fade Listen, a system used within a mixer to allow the operator to listen to a selected signal, regardless of the position of the fader controlling that signal.

Phantom Power

48V DC supply for capacitor microphones, transmitted along the signal cores of a balanced mic cable.

Phase

Timing difference between two electrical waveforms expressed in degrees where 360° corresponds to a delay of exactly one cycle.

Phaser

Effect which combines a signal with a phase-shifted version of itself to produce creative filtering effects. Most phasers are controlled by means of an LFO.

Phono Plug

Hi-fi connector developed by RCA and used extensively on semi-pro, unbalanced recording equipment.

Pickup

Part of a guitar that converts string vibrations to electrical signals.

Pitch

Musical interpretation of an audio frequency.

Pitch Bend

Special control message specifically designed to produce a change in pitch in response to the movement of a pitch bend wheel or lever. Pitch bend data can be recorded and edited, just like any other MIDI controller data, even though it isn't part of the controller message group.

Pitch Shifter

Device for changing the pitch of an audio signal without changing its duration.

Port

Connection for the input or output of data.

Portamento

Gliding effect that allows a sound to change pitch at a gradual rate rather than abruptly when a new key is pressed or MIDI note sent.

Post-Production

Describes work done to a stereo recording after mixing is complete.

Post-Fade

Aux signal taken from after the channel fader so that the aux send level follows any channel fader changes. Normally used for feeding effects devices.

PPM

Peak Programme Meter. A meter designed to register signal peaks rather than the average level.

PPQN
Pulsed Per Quarter Note. Used in the context of MIDI clock-derived sync signals.

PQ Coding
Process for adding pause, cue and other subcode information to a digital master tape in preparation for CD manufacture.

Pre-Fade
Aux signal taken from before the channel fader so that the channel fader has no effect on the aux send level. Normally used for creating foldback or cue mixes.

Preset
Parameter that cannot be altered by the user.

Pressure
Alternative term for aftertouch.

Processor
Device designed to treat an audio signal by changing its dynamics or frequency content. Examples of processors include compressors, gates and equalisers.

PZM

Pressure Zone Microphone. A type of boundary microphone, designed to reject out-of-phase sounds reflected from surfaces within the recording environment.

Q

Measurement of the resonant properties of a filter. The higher the Q, the more resonant the filter and the narrower the range of frequencies that are allowed to pass.

Quantising

Means of moving notes recorded in a MIDI sequencer so that they line up with user-defined subdivisions of a musical bar – 16s, for example. May be used to correct timing errors, but over-quantising can remove the human feel from a performance.

R-DAT

Digital tape machine using a rotating head system.

Real Time

Describes an audio process that can be carried out as the signal is being recorded or played back. The opposite is offline, where the signal is processed in non-real time.

Release

Time taken for a level or gain to return to normal. Often used to describe the rate at which a synthesised sound reduces in level after a key has been released.

Resistance

Opposition to the flow of electrical current. Measured in ohms.

Resonance

Same as Q.

Reverb

Acoustic ambience created by multiple reflections in a confined space.

RF interference

Radio Frequency interference, which is significantly above the range of human hearing.

Ribbon Microphone

Microphone in which the sound-capturing element is a thin metal ribbon suspended in a magnetic filed. When sound causes the ribbon to vibrate, a small electrical current is generated within the ribbon.

Release

Rate at which signal amplitude decays once a key is released.

Resonance
Characteristic of a filter that allows it to selectively pass a narrow range of frequencies. (See *Q*.)

RMS
Root Mean Square. A method of specifying the behaviour of a piece of electrical equipment under continuous sine wave testing conditions.

Roll-off
The rate at which a filter attenuates a signal once it has passed the filter cutoff point.

Sample
Process carried out by an analogue-to-digital converter where the instantaneous amplitude of a signal is measured many times per second (44.1kHz in the case of CD).

Sample
Digitised sound used as a musical sound source in a sampler or additive synthesiser.

Sample And Hold

Usually refers to a feature whereby random values are generated at regular intervals and then used to control another function such as pitch or filter frequency. Sample and hold circuits were also used in old analogue synthesisers to 'remember' the note being played after a key had been released.

Sample Rate
Number of times which an A/D converter samples the incoming waveform each second.

Sawtooth Wave
So called because it resembles the teeth of a saw, this waveform contains only even harmonics.

SCSI
Small Computer System Interface. An interfacing system for using hard drives, scanners, CD-ROM drives and similar peripherals with a computer. Each SCSI device has its own ID number and no two SCSI devices in the same chain must be set to the same number. The last SCSI device in the chain should be terminated either via an internal terminator, where provided, or via a plug-in terminator fitted to a free SCSI socket.

Sequencer
Device for recording and replaying MIDI data, usually

in a multitrack format, allowing complex compositions to be built up a part at a time.

Short Circuit
Low-resistance path that allows electrical current to flow. The term is usually used to describe a current path that exists through a faulty condition.

Sibilance
High-frequency whistling or lisping sound that affects vocal recordings due either to poor mic technique or excessive equalisation.

Side Chain
Part of a circuit that splits off a proportion of the main signal to be processed in some way. Compressors uses aside-chain signal to derive their control signals.

Signal
Electrical representation of input such as sound.

Signal Chain
Route taken by a signal from a system's input to its output.

Signal-To-Noise Ratio
Ratio of maximum signal level to the residual noise, expressed in decibels.

Sine Wave
Waveform of a pure tone with no harmonics.

Single-Ended Noise Reduction
Device for removing or attenuating the noise component of a signal. Doesn't require previous coding, as in the case of Dolby or dbx.

Slave
Device under the control of a master device.

SMPTE
Time code developed for the film industry but now extensively used in music and recording. SMPTE is a real-time code and is related to hours, minutes, seconds and film or video frames rather than to musical tempo.

SPL
Sound-Pressure Level. Measured in decibels.

SPP
Song-Position Pointer (MIDI).

Standard MIDI File
Standard file format that allows MIDI files to be transferred between different sequencers and MIDI file players.

Step Time
System for programming a sequencer in non-real time.

Stereo
Two-channel system feeding left and right loudspeakers.

Stripe
To record time code onto one track of a multitrack tape machine.

Sub-bass
Frequencies below the range of typical monitor loudspeakers. Some define sub-bass as frequencies that can be felt rather than heard.

Subcode
Hidden data within the CD and DAT format that includes such information as the absolute time location, number of tracks, total running time and so on.

Sustain
Part of the ADSR envelope which determines the level to which the sound will settle if a key is held down. Once the key is released, the sound decays at a rate set by the release parameter. Also refers to a guitar's ability to hold notes which decay very slowly.

Sweet Spot

Optimum position for a microphone or a listener relative to monitor loudspeakers.

Sync

System for making two or more pieces of equipment run in synchronism with each other.

Synthesiser

Electronic musical instrument designed to create a wide range of sounds, both imitative and abstract.

Tape Head

Part of a tape machine that transfers magnetic energy to the tape during recording or reads it during playback.

Tempo

Rate of the beat of a piece of music, measured here in beats per minute.

Test Tone

Steady, fixed-level tone recorded onto a multitrack or stereo recording to act as a reference when matching levels.

THD

Total Harmonic Distortion.

Thru
MIDI connector which passes on the signal received at the MIDI In socket.

Timbre
Tonal 'colour' of a sound.

Track
This term dates back to multitrack tape, on which the tracks are physical stripes of recorded material located side by side along the length of the tape.

Transducer
Device for converting one form of energy into another. A microphone is a good example of a transducer, as it converts mechanical energy to electrical energy.

Transparency
Subjective term used to describe audio quality where the high-frequency detail is clear and individual sounds are easy to identify and separate.

Transpose
To shift a musical signal by a fixed number of semitones.

Tremolo
Modulation of the amplitude of a sound using an LFO.

TRS Jack

Stereo-type jack equipped tip, ring and sleeve connections.

Unbalanced

Two-wire electrical signal connection in which the inner (hot or positive) conductor is usually surrounded by the cold (negative) conductor, thus forming a screen against interference.

Unison

To play the same melody using two or more different instruments or voices.

Valve

Vacuum-tube amplification component, also known as a tube.

Velocity

The rate at which a key is depressed. This may be used to control loudness (to simulate the response of instruments such as pianos) or other parameters on later synthesisers.

Vibrato

Pitch modulation using an low-frequency oscillator to modulate a voltage-controlled oscillator.

Voice

Capacity of a synthesiser to play a single musical note. An instrument capable of playing 16 simultaneous notes is said to be a 16-voice instrument.

Volt

Unit of electrical power.

VU Meter

Meter designed to interpret signal levels in roughly the same way as the human ear, which responds more closely to the average levels of sounds rather than to the peak levels.

Wah Pedal

Guitar effects device where a bandpass filter is varied in frequency by means of a pedal control.

Warmth

Subjective term used to describe sound where the bass and low mid frequencies have depth and where the high frequencies are smooth sounding rather than being aggressive or fatiguing. Warm-sounding tube equipment may also exhibit some of the aspects of compression.

Watt

Unit of electrical power.

Waveform

Graphic representation of the way in which a sound wave or electrical wave varies with time.

White Noise

Random signal with an energy distribution that produces the same amount of noise power per Hz.

XG

Yamaha's alternative to Roland's GS system for enhancing the General MIDI protocol so as to provide additional banks of patches and further editing facilities.

XLR

Type of connector commonly used to carry balanced audio signals, including the feeds from microphones.

Y-Lead

Lead split so that one source can feed two destinations. Y-leads may also be used in console insert points, when a stereo jack plug at one end of the lead is split into two monos at the other.